Lecture Notes in Bioinformatics 4366

Edited by S. Istrail, P. Pevzner, and M. Waterman

Editorial Board: A. Apostolico S. Brunak M. Gelfand
T. Lengauer S. Miyano G. Myers M.-F. Sagot D. Sankoff
R. Shamir T. Speed M. Vingron W. Wong

Subseries of Lecture Notes in Computer Science

Karl Tuyls   Ronald Westra
Yvan Saeys   Ann Nowé (Eds.)

Knowledge Discovery and Emergent Complexity in Bioinformatics

First International Workshop, KDECB 2006
Ghent, Belgium, May 10, 2006
Revised Selected Papers

 Springer

Series Editors

Sorin Istrail, Brown University, Providence, RI, USA
Pavel Pevzner, University of California, San Diego, CA, USA
Michael Waterman, University of Southern California, Los Angeles, CA, USA

Volume Editors

Karl Tuyls
Maastricht University, Faculty of Humanities and Science
Maastricht ICT Competence Center, 6200 MD Maastricht, The Netherlands
E-mail: k.tuyls@micc.unimaas.nl

Ronald Westra
Maastricht University, Department of Mathematics
6200 MD Maastricht, The Netherlands
E-mail: westra@math.unimaas.nl

Yvan Saeys
Ghent University
Technologiepark 927, 9052 Ghent, Belgium
E-mail: yvan.saeys@ugent.be

Ann Nowé
Vrije Universiteit Brussel, Faculty of Sciences (WE)
Department of Computer Science, Pleinlaan 2, 1050 Brussels, Belgium
E-mail: ann.nowe@vub.ac.be

Library of Congress Control Number: 2007920820

CR Subject Classification (1998): H.2.8, I.5, J.3, I.2, H.3, F.1-2

LNCS Sublibrary: SL 8 – Bioinformatics

ISSN   0302-9743
ISBN-10   3-540-71036-1 Springer Berlin Heidelberg New York
ISBN-13   978-3-540-71036-3 Springer Berlin Heidelberg New York

Springer is a part of Springer Science+Business Media

springer.com

© Springer-Verlag Berlin Heidelberg 2007
Printed in Germany

Typesetting: Camera-ready by author, data conversion by Scientific Publishing Services, Chennai, India
Printed on acid-free paper   SPIN: 12023929   06/3142   5 4 3 2 1 0

Preface

This book contains selected and revised papers of the International Symposium on Knowledge Discovery and Emergent Complexity in Bioinformatics (KDECB 2006), held at the University of Ghent, Belgium, May 10, 2006.

In February 1943, the Austrian physicist Erwin Schrödinger, one of the founding fathers of quantum mechanics, gave a series of lectures at Trinity College in Dublin titled "What Is Life? The Physical Aspect of the Living Cell and Mind." In these lectures Schrödinger stressed the fundamental differences encountered between observing animate and inanimate matter, and advanced some, at the time, audacious hypotheses about the nature and molecular structure of genes, some ten years before the discoveries of Watson and Crick. Indeed, the rules of living matter, from the molecular level to the level of supraorganic flocking behavior, seem to violate the simple basic interactions found between fundamental particles as electrons and protons. It is as if the organic molecules in the cell 'know' that they are alive. Despite all external stochastic fluctuations and chaos, process and additive noise, this machinery has been ticking for at least 3.8 billion years. Yet, we may safely assume that the laws that govern physics also steer these complex associations of synchronous and seemingly intentional dynamics in the cell. Contrary to the few simple laws that govern the interactions between the few really elementary particles, there are at least tens of thousands of different genes and proteins, with millions of possible interactions, and each of these interactions obeys its own peculiarities. There are different processes involved such as transcription, translation and subsequent folding. How can we ever understand the fundamentals of these complex interactions that emerge from the few empirical observations we are able to make?

The KDECB 2006 Workshop was a great success and provided a forum for the presentation of new ideas and results bearing on the conception of knowledge discovery and emergent complexity in bioinformatics. This event was organized in connection with the 15th Belgium-Netherlands Conference on Machine Learning, held in Ghent, Belgium. The goal of this workshop and this associated book is to increase awareness and interest in knowledge discovery and emergent complexity research in bioinformatics, and to encourage collaboration between machine learning experts, computational biology experts, mathematicians and physicists, so as to give a representative overview of the current state of affairs in this area. Next to a strong program with lectures by leading scientists in this multidisciplinary field, the book contains contributions on how knowledge can be extracted from sophisticated biological systems. Different disciplines, both 'wet' and 'dry,' have contributed to these developments and they will also benefit directly or indirectly from new, intelligent, computational techniques.

Hence, we welcomed scientists and practitioners from several European countries and different scientific areas in Ghent for the 1st Workshop on Knowledge Discovery and Emergent Complexity in Bioinformatics (KDECB 2006).

We hope that our readers will enjoy reading the efforts of the researchers.

Acknowledgements

Organizing a scientific event like KDECB, and editing an associated book, requires the help of many enthusiastic people. First of all, the organizers would like to thank the members of the Program Committee who guaranteed a scientifically strong and interesting LNBI volume. Secondly, we would like to express our appreciation to the invited speakers, Ricardo Grau, Reinhardt Guthke, William H. Majoros, Stan Maree, Grzegorz Rozenberg and Jean-Philippe Vert, for their distinguished contributions to the symposium program. Finally, we would also like to thank the authors of all contributions for submitting their scientific work to the KDECB symposium!

December 2006

Karl Tuyls
Ronald Westra
Yvan Saeys
Ann Nowé

Organization

Organizing Committee

Co-chairs: Karl Tuyls
Ronald Westra
Yvan Saeys
Ann Nowé

Program Committee

Adelmo Cechin
Jeroen Donkers
Reinhard Gutkhe
Thomas Hamelryck
Nicolas Le Novere

Derek Linkens
Bernard Manderick
Ann Nowé
Yvan Saeys
Klaus Stiefel

Elena Tsiporkova
Bram Vanschoenwinkel
Ronald Westra

Table of Contents

Knowledge Discovery and Emergent Complexity in Bioinformatics

Ronald Westra[1], Karl Tuyls[1], Yvan Saeys[2], and Ann Nowé[3]

[1] Department of Mathematics and Computer Science,
Maastricht University and Transnational University of Limburg,
Maastricht, The Netherlands
[2] Department of Plant Systems Biology, Ghent University,
Flanders Interuniversity Institute for Biotechnology (VIB),
Ghent, Belgium
[3] Computational Modeling Lab,
Vrije Universiteit Brussel,
Brussels, Belgium

1 Introduction

In February 1943, the Austrian physicist Erwin Schrödinger, one of the founding fathers of quantum mechanics, gave a series of lectures at the Trinity College in Dublin, entitled "What Is Life? The Physical Aspect of the Living Cell and Mind". In these lectures Schrödinger stressed the fundamental differences encountered between observing animate and inanimate matter, and advanced some at the time audacious hypotheses about the nature and molecular structure of genes, some ten years before the discoveries of Watson and Crick.

Indeed, the rules of living matter, from the molecular level to the level of supraorganic flocking behavior, seem to violate the simple basic interactions found between fundamental particles as electrons and protons. It is as if the organic molecules in the cell 'know' that they are alive. Despite all external stochastic fluctuations and chaos, process and additive noise, this machinery is ticking for at least 3.8 billion years. Yet, we may safely assume that the laws that govern physics also steer these complex associations of synchronous and seemingly intentional dynamics in the cell. Contrary to the few simple laws that govern the interactions between the few really elementary particles, there are at least tens of thousands of different genes and proteins, with millions of possible interactions, and each of these interactions obeys its own peculiarities. There are different processes involved like transcription, translation and subsequent folding. How can we ever understand the fundamentals of these complex interactions that emerge from the few empirical observations we are able to make.

The KDECB 2006 Symposium, and this associated book, is intended to provide a forum for the presentation of new ideas and results bearing on the conception of knowledge discovery and emergent complexity in bioinformatics. The goal of this symposium is to increase awareness and interest in knowledge discovery and emergent complexity research in Bioinformatics, and encourage collaboration between Machine Learning experts, Computational Biology experts,

K. Tuyls et al. (Eds.): KDECB 2006, LNBI 4366, pp. 1–9, 2007.
© Springer-Verlag Berlin Heidelberg 2007

Mathematicians and Physicists, and give a representative overview of the current state of affairs in this area. Next to a strong program with lectures of leading scientists in this multi-disciplinary field, we present contributions that cover on how knowledge can be extracted from, and complexity emerges in sophisticated biological systems. Different disciplines, both 'wet' and 'dry', have contributed to these developments and they will also benefit directly or indirectly from new, intelligent, computational techniques.

In the remainder of this document the three main themes of this book are introduced and discussed, namely, (i) Machine Learning for Bioinformatics, (ii) Mathematical modeling of gene-protein networks, and, (iii) Nature-inspired computation.

2   Machine Learning for Bioinformatics

During the past decades, advances in genomics have generated a wealth of biological data, increasing the discrepancy between what is observed and what is actually known about life's organisation at the molecular level. To gain a deeper understanding of the processes underlying the observed data, pattern recognition techniques play an essential role.

The notion of a *pattern* however, needs to be interpreted in a very broad sense. Essentially, we could define a pattern as everything that is the opposite of chaos. Thus the notion of *organisation* can be associated with a pattern. The goal of pattern recognition techniques then is to elucidate the organisation of the pattern, resulting in a wide range of subtasks such as recognition, description, classification, and grouping of patterns.

In bioinformatics, techniques to learn the theory *automatically* from the data (machine learning techniques) play a crucial role, as they are a first step towards interpreting the large amounts of data, and extracting useful biological knowledge from it. Machine learning techniques are generally applied for the following problems: classification, clustering, construction of probabilistic graphical models, and optimisation.

In classification (sometimes also referred to as supervised learning) the goal is to divide objects into classes, based on the characteristics of the objects. The rule that is used to assign an object to a particular class is termed the classification function, classification model, or classifier. Many problems in bioinformatics can be cast into a classification problem, and well established methods can then be used to solve the task. Examples include the prediction of gene structures [4,26,37], which often is the first step towards a more detailed analysis of the organism, the classification of microarray data [17,21], and recently also classification problems related to text mining in biomedical literature [23]. The computational gene prediction problem is the problem of the automatic annotation of the location, structure, and functional class of protein-coding genes. A correct annotation forms the basis of many subsequent modeling steps, and thus should be done with great care. Driven by the explosion of genome data, computational approaches to identify genes have thus proliferated, thereby depending strongly on machine learning techniques.

A second class of problems in bioinformatics concerns the topic of clustering, also termed unsupervised learning, because no class information is known a priori. The goal of clustering is to find natural groups of objects (clusters) in the data that is being modeled, where objects in one cluster should be similar to each other, while being at the same time different from the objects in another cluster. The most common examples of clustering in bioinformatics concern the clustering of microarray expression data [10,19,39], and the grouping of sequences, e.g. to build phylogenetic trees [13].

Probabilistic graphical models [31] have proliferated as a useful set of techniques for a wide range of problems where dependencies between variables (objects) need to be modeled. Formally, they represent multivariate joint probability densities via a product of terms, each of which only involves a few variables. The structure of the problem is then modeled using a graph that represents the relations between the variables, which allows to reason about the properties entailed by the product. Common applications include the inference of genetic networks in systems biology [38] and Bayesian methods for constructing phylogenetic trees [34]. Other examples of applications of machine learning techniques in bioinformatics include the prediction of protein structure (which can be cast into an optimisation problem), motif identification in sequences, and the combination of different sources of evidence for analysis of global properties of bio(chemical) networks. In all of these domains, machine learning techniques have proven their value, and new methods are constantly being developed [25].

3   Modeling the Interactions Between Genes and Proteins

A prerequisite for the successful reconstruction of gene-protein networks is the way in which the dynamics of their interactions is modeled. The formal mathematical modeling of these interactions is an emerging field where an array of approaches are being attempted, all with their own problems and short-comings. The underlying physical and chemical processes involved are multifarious and hugely complex. This condition contrasts sharply with the modeling of inanimate Nature by physics. While in physics huge quantities of only a small amount of basic types of elementary particles interact in a uniform and deterministic way provided by the fundamental laws of Nature, the situation in gene-protein interactions deals with tens of thousands of genes and possibly some million proteins. The quantities thereby involved in the actual interactions are normally very small, as one single protein may be able to (in)activate a specific gene, and thereby change the global state of the system. For this reason, gene regulatory systems are much more prone to stochastic fluctuations than the interactions involved in normal inorganic reactions. Moreover, each of these interactions is different and involves its own peculiar geometrical and electrostatic details. There are different processes involved like transcription, translation and subsequent folding. Therefore, the emergent complexity resulting from gene regulatory networks is much more difficult to comprehend.

In the past few decades a number of different formalisms for modeling the interactions amongst genes and proteins have been presented. Some authors focus on specific detailed processes such as the circadian rhythms in *Drosophila* and *Neurospora* [16,18], or the cell cycle in *Schizosaccharomyces* (Fission yeast) [30]. Others try to provide a general platform for modeling the interactions between genes and proteins. For a thorough overview consult de Jong (2002) in [6], Bower (2001) in [3], and others [12,14,20].

Traditionally, much emphasis lay on static models, where the relations between genes and proteins are considered fixed in time. This was in line with the impressive developments in microarray technology that opened a window towards reconstructing static genetic and metabolic pathways, as for instance demonstrated in [36]. Successful static models are the Logical Boolean networks consult [2,3,5,1], and on Bayesian Networks consult [14,40,41]. In discrete event simulation models the detailed biochemical interactions are studied. Considering a large number of constituents, the approach aims to derive macroscopic quantities. More information on discrete event modeling can be found in[3].

In contrast to the static networks, the aim in modeling dynamic networks is to explain the macroscopic network complexity from the molecular dynamics and reaction kinetics. The approach to modeling the dynamical interactions amongst genes and proteins is by considering them as biochemical reactions, and thus representing them as traditional 'rate equations'. The concept of chemical rate equations, dating back to Van 't Hoff, consists of a set of differential equations, expressing the time derivative of the concentration of each constituent of the reaction as some rational function of the concentrations of all the constituents involved. In general, the syntax of the chemical reactions is mapped on the syntax of the rate equations, as e.g. in the Michaelis-Menten equation for enzyme kinetics. More on the physical basis of rate equations can be found in [48].

Though the truth of the underlying biochemical interactions between the constituents is generally accepted, the idea of representing them by rate equations involves a number of major problems. First of all, the rate equation is not a fundamental law of Nature like the great conservation laws of Energy and Momentum, but a statistical average over the entire ensemble of possible microscopic interactions. The applicability of the rate equation therefore relates to the law of large numbers. In normal inorganic reactions this requirement holds. However, in inhomogeneous mixtures or in fast reactions the actual dynamics will depart significantly from this average. Also in case of gene-, RNA-, and protein-interactions this condition will not hold as we will discuss later. Second, the Maxwell velocity distribution should apply, otherwise the collision frequency between the constituents would not be proportional to their concentrations, and details of the velocity distribution would enter. This condition is met easily in the presence of a solvent or an inert gas, but difficult to attain for macromolecules in a cytoplasm. The same holds for the distribution of the internal degrees of freedom of the constituents involved, such as rotational and vibrational energies. The distribution of their energies should have the same 'temperature' as in the Maxwell velocity distribution, otherwise this would affect the rate of the collisions that result in an actual chemical reaction. Also this condition is not easily met by gene-protein interactions. Finally, the temperature of the reaction should be constant in space and time - this condition may be accounted for in this context.

So, rate equations are statistical approximations that hold under above requirements. Under these conditions they predict the average number of reactive collisions. The actual observed number will fluctuate around this number, depending on the details of the microscopic processes involved. In case of biochemical interactions between genes and proteins at least some of the conditions will be violated and therefore the applicability of the concept of rate equations is valid only for genes with sufficient high transcription rates. This is confirmed by recent experimental findings by Swain and Elowitz [11], [35], [42], [43].

Dynamic gene-protein networks can lead to mathematical complexities in modeling and identification [27,28,8]. To overcome these problems, some authors have proposed to model them as piecewise linear models, as introduced by Glass and Kauffman [15]. Such models can be demonstrated to be memory-bounded Turing-machines [2]. de Jong *et al.* [6,7] have proposed qualitative piecewise linear models rather than a quantitative models, because the correct underlying multifarious mathematical expressions are not tractable. In spite of the intuitive attractiveness of this idea, there are a number of conceptual and practical problems in applying these techniques in practical situations. In biology piecewise linear behaviour is frequently observed, as for instance in embryonic growth where the organism develops by transitions through a number of well-defined 'check points'. Within each such checkpoint the system is in relative equilibrium. However, it should be mentioned that there is an ongoing debate on the modeling of gene-protein dynamics as *checkpoint mechanisms* versus *limit-cycle oscillators*, see [33,44].

Others have employed specific characteristics of the networks to construct a viable reconstruction algorithm, such as the sparsity and hierarchy in the network interactions [8,49,32].

4   Nature-Inspired Computing

In the sections above, we gave an overview of approaches and techniques from computer science and mathematics that are promising in order to model biological phenomena such as gene networks, protein structure, etc. We can however go one step further, and try to model the emergent collective intelligence, arising in nature from local, simple interactions between simple units, which can be biological cells, neurons as well as insects as ants and bees. Using insights from how this complexity and global intelligence emerges in nature, we can develop new computational algorithms to solve hard problems. Well known examples are Neural Networks and Genetic Algorithms. Whereas Neural Networks are inspired on the working of the brain, Genetic Algorithms are based on the model of natural evolution. Another natured inspired technique is reinforcement learning. Reinforcement learning [22,45] finds its roots in animal learning. It is well known that, by operand or instrumental conditioning, we can teach an animal to respond in some desired way. The learning is done by rewarding and punishing the learner appropriately, and as a result the likelihood of the desired behaviour is increased during the learning process, whereas undesired behaviour will become less likely.

The objective of a reinforcement learner is to discover a policy, meaning a mapping from situations to actions, so as to maximise the reinforcement it receives. The reinforcement is a scalar value which is usually negative to express a punishment, and positive to indicate a reward. Unlike supervised learning techniques, reinforcement learning methods do not assume the presence of a teacher who is able to judge the action taken in a particular situation. Instead the learner finds out what the best actions are by trying them out and by evaluating the consequences of the actions by itself. For many problems, such as planning problems, the consequences of the action are not immediately apparent after performing the action, but only after a number of other actions have been taken. In other words the selected action may not only affect the immediate reward/punishment the learner receives, but also the reinforcement it might get in subsequent situations, i.e. the delayed rewards or punishments. Reinforcement learning techniques such as Q-learning and Adaptive Critique techniques, can deal with this credit assignment problem and are guaranteed to converge to an optimal policy, as long as some conditions, such as the environment experienced by the learner should be Markovian and the learner should be allowed sufficient exploration, are met.

More recently other nature inspired techniques such as Ant Colony Optimisation (ACO) [9] received a lot of attention. ACO techniques are inspired by the behaviour of ants. It is well known that one single ant on its own cannot do anything useful, but a colony of ants is capable of performing complex behaviour. The complex behaviour emerges due to the fact that ants can communicate indirectly with each other, by laying a pheromone trail in the environment. This pheromone signal can be observed by other ants, and this will influence their own behaviour. The more pheromone is sensed by an ant in some direction, the more it will be attracted in that direction, and the more the pheromone will be reinforced. ACO algorithms have been successfully applied to complex graph problems such as large instances of the travelling salesman problem. ACO techniques are closely related to the Reinforcement Learning technique mentioned in the previous paragraph, however they do not come with straightforward convergence proofs. As is illustrated in [46] by Verbeeck *et al.* it is possible to provide a clean proof of convergence by expressing the mapping between the ACO pheromone updating mechanism and interconnecting learning automata [29]. The insight into the convergence issues of these algorithms is crucial in order to have a wider acceptance of these techniques.

Recent investigations [24,47] have also opened up the possibility of applying recruitment and navigational techniques from honeybees to computational problems as for instance foraging. Honeybees use a strategy named Path Integration. By employing this strategy, bees always know a direct path towards their destination and their home. Bees employ a direct recruitment strategy by dancing in the nest. Their dance communicates distance and direction towards a destination. Ants, on the other hand, employ an indirect recruitment strategy by accumulating pheromone trails. When a trail is strong enough, other ants are attracted to it and will follow this trail towards a destination. Both strategies provide the insects with an efficient way of foraging.

References

1. Arkin A., Ross J., McAdams H.H. (1994), Computational functions in biochemical reaction networks. *Biophys. Journal*, Vol. 67, pp. 560–578.
2. Ben-Hur A., Siegelmann H.T. (2004), Computation in Gene Networks. *Chaos*, Vol. 14(1) pp. 145–151.
3. Bower J.M., Bolouri H.(Editors) (2001), Computational Modeling of Genetic and Biochemical Networks. *MIT Press*, 2001.
4. Burge, C., Karlin, S. (1997), Prediction of complete gene structures in human genomic DNA. *J. Mol. Biol*, Vol. 268, pp. 78–94.
5. Davidson E.H. (1999), A View from the Genome: Spatial Control of Transcription in Sea Urchin Development, *Current Opinions in Genetics and Development*, Vol. 9, pp. 530–541.
6. de Jong H. (2002), Modeling and Simulation of Genetic RegulatorySystems: A Literature Review. *Journal of Computational Biology*, Vol. 9(1), pp. 67–103.
7. de Jong H., Gouze J.L., Hernandez C., Page M., Sari T., Geiselmann J. (2004), Qualitative simulation of genetic regulatory networks using piecewise-linear models. *Bull Math Biol.*, Vol. 66(2), pp. 301–40.
8. D'haeseleer P., Liang S., Somogyi R. (2000), Genetic Network Inference: From Co-Expression Clustering to Reverse Engineering. *Bioinformatics*, Vol. 16(8), pp. 707–726.
9. Dorigo M. and Sttzle T. (2004), Ant Colony Optimization. MIT Press. (2004).
10. Eisen, M. B., Spellman, P. T., Brown, P. O., Botstein, D. (1998), Cluster analysis and display of genome-wide expression patterns. *Proc. Natl. Acad. Sci. USA*, Vol. 95(25), pp 14863-14868
11. Elowitz M.B., Levine A.J., Siggia E.D., Swain P.S. (2002), Stochastic gene expression in a single cell. *Science*, Vol. 297, pp. 1183–1186.
12. Endy, D, Brent, R. (2001), Modeling Cellular Behavior. *Nature*, Vol. 409(6818), pp. 391–395.
13. Felsenstein, J. (2004), Inferring Phylogenies. Sinauer Associates, Sunderland, Mass.
14. Friedman, N., Linial, M., Nachman, I., Pe'er, D. (2000), Using Bayesian Networks to analyze expression data. *Journal of Computational Biology*, Vol. 7, pp. 601–620.
15. Glass L., Kauffman S.A. (1973), The Logical Analysis of Continuous Non-linear Biochemical Control Networks, *J.Theor.Biol.*, Vol. 39(1), pp. 103–129
16. Goldbeter A (2002), Computational approaches to cellular rhythms. *Nature*, Vol. 420, pp. 238–45.
17. Golub, T., Slonim, D., Tamayo, P., Huard, C., Gaasenbeek, M., Mesirov, J., Coller, H., Loh, M., Downing, J., Caligiuri, M. (1999), Molecular Classification of Cancer: Class Discovery and Class Prediction by Gene Expression Monitoring. *Science*, Vol. 286, pp. 531–537.

18. Gonze D, Halloy J, and Goldbeter A (2004), Stochastic models for circadian oscillations : Emergence of a biological rhythm. *Int J Quantum Chem*, Vol. 98, pp. 228–238.
19. Hastie, T., Tibshirani, R., Eisen, M. B., Alizadeh, A., Levy, R., Staudt, L., Chan, W. C., Botstein, D., Brown, P. (2000), Gene shaving as a method for identifying distinct sets of genes with similar expression patterns. *Genome Biol.*, Vol. 1(2),research0003.10003.21.
20. Hasty J., McMillen D., Isaacs F., Collins J. J., (2001), Computational studies of gene regulatory networks: in numero molecular biology. *Nature Reviews Genetics*, Vol. 2(4), pp. 268– 279.
21. Inza, I., Larrañaga, P., Blanco, R., Cerrolaza, A.J. (2004), Filter versus wrapper gene selection approaches in DNA microarray domains. *Artificial Intelligence in Medicine*, special issue in "Data mining in genomics and proteomics", Vol.31(2), pp 91–103.
22. Kaelbling L.P., Littman L.M. and Moore A.W. (1996), Reinforcement learning: a survey, Journal of Artificial Intelligence Research, 4 (1996) 237-285.
23. Krallinger, M., Valencia, A. (2005), Text-mining and information-retrieval services for molecular biology. *Genome Biol.*, Vol. 6(7), 224
24. Lambrinos, D., Moller, R., Labhart, T., Pfeifer, R., and Wehner, R. (2000). A mobile robot employing insect strategies for navigation. Robotics and Autonomous Systems, Vol. 30, Nos. 12, pp. 3964.
25. Larrañaga, P., Calvo, B., Santana, R., Bielza, C., Galdiano, J., Inza, I., Lozano, J.A., Armañanzas, R., Santafé, R., Pérez, A., Robles, V. (2006), Machine Learning in Bioinformatics. *Briefings in Bioinformatics*, Vol.7(1), pp. 86–112.
26. Mathé, C., Sagot, M.F., Schiex, T. and Rouzé, P. (2002), Current methods of gene prediction, their strengths and weaknesses. *Nucleic Acids Res.*, Vol. 30(19), pp. 4103–17.
27. Mestl T., Plahte E., Omholt S.W. (1995*a*), A Mathematical Framework for describing and analysing and Analyzing Gene Regulatory Networks. *J. Theor. Biol.*, Vol. 176(2), pp. 291–300.
28. Mestl T., Plahte E., Omholt S.W. (1995*b*), Periodic Solutions in Systems of Piecewise-Linear Systems. *Synamic Stability of Systems*, Vol. 10(2), pp. 179–193.
29. Narendra K. and Thathachar M., Learning Automata: An Introduction, Prentice-Hall International, Inc, (1989).
30. Novak B, Tyson JJ (1997), Modeling the control of DNA replication in fission yeast. *Proc. Natl. Acad. Sci. USA*, Vol. 94, pp. 9147–9152.
31. Pearl, J. (1988), Probabilistic reasoning in intelligent systems: networks of plausible inference. Morgan Kaufmann Publishers, 1988.
32. Peeters R.L.M., Westra R.L. (2004), On the identification of sparse gene regulatory networks, *Proc. of the 16th Intern. Symp. on Mathematical Theory of Networks and Systems* (MTNS2004) Leuven, Belgium July 5-9, 2004
33. Rao, C.V., Wolf, D.M., Arkin, A.P. (2002), Control, exploitation and tolerance of intracellular noise. *Nature*, Vol. 420, pp. 231–237.
34. Ronquist, F., J. P. Huelsenbeck (2003), MRBAYES 3: Bayesian phylogenetic inference under mixed models. *Bioinformatics*, Vol. 19, pp. 1572–1574.
35. Rosenfeld, N., Young, J.W., Alon, U., Swain, P.S., Elowitz, M.B., Gene regulation at the single-cell level. *Science*, Vol. 307, pp. 1962.
36. Rustici, G., Mata, J., Kivinen, K., Lio, P., Penkett, C.J., Burns, G., Hayles, J., Brazma, A., Nurse, P., Bahler, J. (2004), Periodic gene expression program of the fission yeast cell cycle. *Nature Genetics*, Vol. 36(8), pp. 809–17.

37. Salzberg, S.L., Pertea, M., Delcher, A.L., Gardner, M.J. and Tettelin, H. (1999), Interpolated Markov models for eukaryotic gene finding. *Genomics*, Vol. 59, pp. 24–31.
38. Segal, E., Shapira, M., Regev, A., Pe'er, D., Botstein, D., Koller, D., Friedman, N. (2003), Module Networks: Identifying Regulatory Modules and their Condition Specific Regulators from Gene Expression Data. *Nature Genetics*, Vol. 34(2), pp. 166–76.
39. Sheng, Q., Moreau, Y., De Moor, B. (2003), Biclustering microarray data by Gibbs sampling. *Bioinformatics*, Vol. 19 (Suppl. 2), pp. ii196-ii205.
40. Smith, V. A., Jarvis, E. D., Hartemink, A. J. (2002), Evaluating Functional Network Inference Using Simulations of Complex Biological Systems. *Proc. of the 10th international conference on Intelligent Systems for Molecular Biology*.
41. Somogyi R., Fuhrman S., Askenazi M., Wuensche A. (1997), The Gene Expression Matrix: Towards the Extraction of Genetic Network Architectures. Nonlinear Analysis. *Proc. of Second World Cong. of Nonlinear Analysis* (WCNA96), Vol.30(3) pp. 1815–1824.
42. Swain P.S. (2004), Efficient attenuation of stochasticity in gene expression through post-transcriptional control. *J. Mol. Biol.*, Vol. 344, pp. 965.
43. Swain, P.S., Elowitz, M.B., Siggia. E.D. (2002), Intrinsic and extrinsic contributions to stochasticity in gene expression. *Proc. Natl. Acad. Sci. USA*, Vol. 99(20), pp.12795–800.
44. Steuer R. (2004), Effects of stochasticity in models of the cell cycle:from quantized cycle times to noise-induced oscillations. *Journal of Theoretical Biology*, Vol. 228, pp. 293–301.
45. Sutton, R.S., Barto, A.G. : Reinforcement Learning: An introduction. Cambridge, MA: MIT Press (1998).
46. Verbeeck K. and Nowé A., Colonies of Learning Automata, IEEE Transactions on Systems, Man and Cybernetics - Part B, Special Issue on Learning Automata: Theory, Paradigms and Applications, 32 (2002) 772-780.
47. Wolf, H. and Wehner, R. (2000). Pinpointing food sources: olfactory and anemotactic orientation in desert ants, Cataglyphis Fortis. The Journal of Experimental Biology, Vol. 203, pp. 857868.
48. van Kampen N. G. (1992), Stochastic Processes in Physics and Chemistry, Elsevier ScienceB. V., Amsterdam, (1992).
49. Yeung M.K.S., Tegnér J., Collins J.J. (2002), Reverse engineering gene networks using singular value decomposition and robust regression, *Proc. Natl. Acad. Sci. USA*, Vol. 99(9), pp. 6163–6168.

Boolean Algebraic Structures of the Genetic Code: Possibilities of Applications

Ricardo Grau[1], Maria del C. Chavez[1], Robersy Sanchez[2], Eberto Morgado[1], Gladys Casas[1], and Isis Bonet[1]

[1] Center of Studies on Informatics, Central University of Las Villas, Santa Clara, CP 54830, Cuba
[2] Research Institute of Tropical Roots, Tuber Crops and Banana (INIVIT), Biotechnology Group, Santo Domingo, Cuba
{Rgrau,MChavez,Robersy,Morgado,Gladita,Ibonet}@uclv.edu.cu

Abstract. Authors had reported before two dual Boolean algebras to understand the underlying logic of the genetic code structure. In such Boolean structures, deductions have physico-chemical meaning. We summarize here that these algebraic structures can help us to describe the gene evolution process. Particularly in the experimental confrontation, it was found that most of the mutations of several proteins correspond to deductions in these algebras and they have a small Hamming distance related to their respective wild type. Two applications of the corresponding codification system in bioinformatics problems are also shown. The first one is the classification of mutations in a protein. The other one is related with the problem of detecting donors and acceptors in DNA sequences. Besides, pure mathematical models, Statistical techniques (Decision Trees) and Artificial Intelligence techniques (Bayesian Networks) were used in order to show how to accomplish them to solve these knowledge-discovery practical problems.

Keywords: Genetic code; Boolean algebra; mutant sequence analysis; splice site prediction; decision trees; Bayesian networks.

1 Introduction

The non-random organization of the genetic code has been pointed out and the Manifold hypothesis was proposed to explain the enigmatic order observed [1], [2], [3], [4]. We summarize in this introduction the results published by the authors about the order determined from two Boolean algebraic structures for the set of the four bases and extended to the codon set [5], [6], [7] and we show in the paper some physico-chemical interpretations and their possibilities of applications.

1.1 The Boolean Algebras in the Set of Four Bases

In every Boolean algebra X with operators: $\wedge$ (AND), $\vee$ (OR) and $\neg$ (NOT), for any two elements α, $\beta \in$ X we have $\alpha \leq \beta$, if and only if $\neg \alpha \vee \beta = 1$ (1 is the neutral element for the operation "$\wedge$", as 0 is the neutral element for the operation "$\vee$"). If $\neg \alpha \vee \beta = 1$ it

K. Tuyls et al. (Eds.): KDECB 2006, LNBI 4366, pp. 10–21, 2007.

is also said that β is deduced from α. Furthermore, if α≤β and β≤α the elements α and β are said to be comparable. Otherwise, they are said not to be comparable. The last partial order is defined by using the hydrogen bond number and the chemical types of purines {A, G} and pyrimidines {U, C} bases. Next, the Boolean lattice is built on the base triplet set (64-codon set) that will be the direct third power of the Boolean lattice of four DNA bases.

The Boolean lattice of the four bases is built assuming that the complementary bases (in the Boolean sense) are the complementary bases in the DNA molecule. Furthermore every Boolean lattice with 4 elements must have two non-comparable elements, a maximum, a minimum and two non-comparable elements. At this point we assumed that the maximum element in the Boolean lattice of the genetic code will be the direct third power of the maximum element in the Boolean lattice of the bases and we come to the correspondence: A→AAA, C→CCC, U→UUU, G→GGG.

1) Both codons GGG and CCC have the same maximum hydrogen bond numbers. This property is reflected in the Boolean lattice, so that the GGG complementary element has to be CCC. Furthermore, both codons code for small amino acid side chains with small polarity difference Glycine and Proline respectively. Then this similar property determines that these elements are comparable.

2) Both codons UUU and AAA have the same minimum hydrogen bond numbers and then, the complementary element in the Boolean lattice of UUU has to be AAA. But, these codons respectively code for amino acid side chains with extreme opposite polarities, Leucine (a hydrophobic residue) and Lysine (having a strong polar group). Consequently, this property suggests these elements are not to be comparable.

These last observations allow us to choose two dual Boolean lattices of the four bases that will be conventionally called Primal and Dual Boolean lattices. From the first observation it might think that the maximum element in the primal Boolean lattice is C and the minimum element is G; or that in the dual Boolean lattice, the maximum element is G and the minimum element is C. The second observation means that the elements U and A are not comparable and therefore, they should not be the maximum or minimum elements in a Boolean lattice with biological meaning. So, we have two Boolean lattices (B(X), ∨ , ∧) (primal lattice) and (B'(X), ∧ , ∨) (dual lattice), where X={U, C, G, A}. The Hasse diagrams of the two duals Boolean lattices obtained are shown in Fig.1.

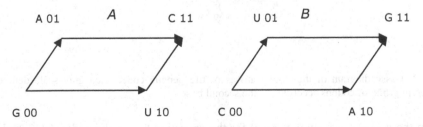

Fig. 1. The Hasse diagrams of the four bases Boolean lattices. *A*: Primal and *B*: Dual.

It is obvious that the primal and dual terms in these Boolean lattices will be interchanged and as we will see later, they do not affect the biological meaning. To simplify the notation we will refer simultaneously to both algebras as B(X).

These Boolean lattices have their equivalent Boolean algebras. From the algebraic point of view B(X) is isomorphic to $(Z_2)^2$ with the classical binary operations ($Z_2=\{0,1\}$). This isomorphism is derived from the fact that all Boolean lattices with the same number of elements are isomorphic. Then, it is possible to represent the primal lattice by means of the correspondence: G$\leftrightarrow$ 00; A$\leftrightarrow$ 01; U$\leftrightarrow$ 10; C$\leftrightarrow$ 11. Likewise, for the dual lattice we have: C$\leftrightarrow$ 00; U$\leftrightarrow$ 01; A$\leftrightarrow$ 10; G$\leftrightarrow$ 11.

1.2 The Boolean Algebra in the Set of Codons

The Boolean algebras of codons are obtained from the direct third power of the Boolean algebras B(X) of the four DNA bases. Explicitly, the direct product C=B(X)xB(X)xB(X) is taken as the Boolean algebras of codons. These algebras are isomorphic to $((Z_2)^6, \vee, \wedge)$, induced by the isomorphism B(X) $\leftrightarrow(Z_2)^2$, so, for instance (in the primal algebra): GUC $\vee$ CAG = CCC $\leftrightarrow$ 001011$\vee$ 110100=111111, GUC $\wedge$ CAG = GGG $\leftrightarrow$ 001011 $\wedge$ 110100=000000, $\neg$(GUC) = CAG $\leftrightarrow$ $\neg$(001011)=110100. Thus, we start from the source alphabet of the genetic code, consisting of the four nucleotides of the DNA and the mRNA and arrive at the second extension of that alphabet with 2^6=64 letter-codons of the genetic code and we represent the Hasse diagram of both lattices (primal and dual) in the Fig. 2.

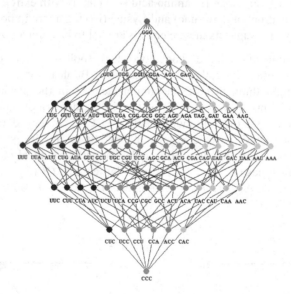

Fig. 2. Hasse diagram of the Boole lattices of the genetic code. Each gray scale denotes a different group of codons according to the second base.

In the primal the neutral element for the operation "$\vee$" is the codon GGG and for the operation "$\wedge$" is the codon CCC. In both lattices, codons are read in the 5′$\rightarrow$3′ direction and anticodon in the 3′$\rightarrow$5′ direction following the standard convention.

Consequently the anticodon of the codon $^{5'}GUC^{3'}$ represented by 001011 in the primal lattice, is the triplet $^{3'}CAG^{5'}$ similarly represented by 001011 in the dual lattice or represented by $110100=\neg(001011)$ in the primal lattice.

In the primal lattice the codon $Y_1Y_2Y_3$ is deduced from the codon $X_1X_2X_3$ if and only if $\neg(X_1X_2X_3)\lor Y_1Y_2Y_3=CCC$. We draw the Hasse diagram lines without arrows to represent deductions in both senses (primal and dual), and so, paths connect comparable codons. Particularly, codons with U as a second base will appear in chains (path of maximum length) where codons with A as a second base will not.

The Hamming distance (d_H) between two codons, represented as binary sextuplets, corresponds to the number of different digits between them. That is, for instance $d_H(GUC,CAG) = d_H(001011,110100)=6$ represents the distance between nodes in the Hasse diagram that is shown in Fig. 2.

Now it's possible to extend these ideas to the set of sequence of the same length N (with N codons). We can operate "logically" with them as we do in $((Z_2)^6)^N$ and we can speak about deductions and Hamming distance between these genes too. It will be important for the study of mutations as we show in the following epigraph.

2 Some Physico-chemical Interpretations and Experimental Confrontations

At this point, we want to show some model connections with the experimental data to help in the understanding of the mutation process of molecular evolution. At first, it was already observed than codons that code to amino acids with extreme hydrophobic difference are in different chains with maximum length in the lattice structure.

2.1 Boolean Deductions

Both Boolean algebras reflect the well known experimental results: single-base substitutions are strongly conservative in regard to amino acid changes in polarity. It has been pointed out that the genetic code reduces the effects of point mutations and minimizes subsequent transcriptions and translations errors to make possible the reproduction of genetic information [1], [3], [8], [9]. Next, it should be expected that the most frequently observed mutations minimize the above effect in proteins. Hence, the fitting of our model with the experimental fact implies that the most frequent mutation in protein should be deducted from the respective wild type.

Authors have calculated the number of deductions in a set of the 749 HIV-1 protease mutant genes deduced from the respective gene of the HXB2 strain. Each protease sequence was aligned with the wild type. The calculations showed that from 11 182 mutations, 88.3% are comparable. Moreover if only those mutations in HIV-1 protease gene conferring drug resistance were taken into account for HIV-1 protease, one can see that only a few mutations are not comparable [5]. Similar results were obtained with other proteins. In Table 1, we show a sample of analyzed mutations in HIV-1 reverse transcriptase that confer drug resistance. One can see that only a small number of mutations are not comparable.

Table 1. The deductible mutations found in the HIV reverse transcriptase gene conferring drug resistance. Most of the reported mutations in HIV reverse transcriptase gene are comparables. Mutations that are not comparables are presented in bold face. In the table, there are only single point mutations, but there have been sequential mutations reported in different combinations.

Amino acid changes	Codon Mutation	Antiviral*	Amino acid changes	Codon Mutation	Antiviral*
A 62 V	GCC -> GTC	Multi-drug resistant	H 208 Y	CAT -> TAT	AZT, lamivudine, PFA
A 98 G	GCA -> GGA	L-697,661	I 135 M	ATA -> ATG	Delavirdine, BI-RG-587
D 67 A	GAC -> GCC	AZT (zidovudine)	I 135 T	ATA -> ACA	Delavirdine, nevirapine
D 67 E	GAC -> GAG	Multi-drug resistant	K 101 E	AAA -> GAA	Multi-drug resistant
D 67 G	GAC -> GAG	(+)dOTFC	K 101 Q	AAA -> CAA	LY-300046 HCl
D 67 G	GAC -> GGC	Multi-drug resistant	K 103 N	AAA -> AAC	Multi-drug resistant
D 67 N	GAC -> AAC	AZT (zidovudine)	K 103 R	AAA -> AGA	LY-300046 HCl, I-EBU
E 138 A	GAG -> GCG	TSAO	K 103 T	AAA -> ACA	S-1153, UC-42
E 138 K	GAG -> AAG	Multi-drug resistant	K 70 E	AAA -> GAA	3TC, PMEA
E 44 A	GAA -> GCA	3TC (lamivudine)	K 70 R	AAA -> AGA	AZT (zidovudine)
E 44 D	GAA -> GAC	3TC (lamivudine)	K 70 S	AAA -> AGA	ddI, d4T
E 89 G	GAA -> GGA	PFA (foscarnet)	**L 100 I**	**TTA -> ATA**	Multi-drug resistant
E 89 K	GAA -> GGA	PFA (foscarnet)	L 210 W	TTG -> TGG	AZT, lamivudine, PFA
F 116 Y	**TTT -> TAT**	Multi-drug resistant	L 214 F	CTT -> TTT	AZT, ph-AZT
F 77 L	TTC -> CTC	Multi-drug resistant	L 74 V	TTA -> GTA	1592U89 (abacavir)
G 141 E	GGG -> GAG	UC-16	P 119 S	CCC -> TCC	F-ddA (lodenosine)
G 190 A	GGA -> GCA	BI-RG-587,	P 157 S	CCA -> TCA	3TC (lamivudine)
G 190 E	GGA -> GAA	Multi-drug resistant	**Q 145 M**	**CAG -> ATG**	Multi-drug resistant
G 190 Q	GGA -> CAA	Multi-drug resistant	**Q 151 M**	**CAG -> ATG**	Multi-drug resistant
G 190 S	GGA -> TCA	DMP-266 (efavirenz)	**Q 161 L**	**CAA -> CTA**	PFA (foscarnet)
G 190 T	GGA -> ACA	DMP-266 HBY 097	R 211 K	AGG -> AAG	AZT , lamivudine
G 190 V	GGA -> GTA	DMP-266 BI-RG-587	S 156 A	TCA -> GCA	PFA (foscarnet)
G 190 V	GGA -> GTA	DMP-266 (efavirenz)	T 139 I	ACA -> ATA	ADAMII, Calanolide A
G 190 V	GGA -> GTA	BI-RG-587	V 106 A	GTA -> GCA	Multi-drug resistant
H 208 Y	CAT -> TAT	AZT, lamivudine,	V 108 I	GTA -> ATA	DMP-266, trovirdine
I 135 M	ATA -> ATG	Delavirdine	V 118 I	GTT -> ATT	3TC (lamivudine)
I 135 T	ATA -> ACA	Delavirdine	**V 179 D**	**GTT -> GAT**	Multi-drug resistant
K 101 Q	AAA -> CAA	LY-300046 HC1	V 179 F	GTT -> TTT	TMC125

Remark 1. All information on mutations contained in this printed table was taken from the Los Alamos web site: http://resdb.lanl.gov/Resist DB.

Authors have shown that in the beta-globin gene the most frequent mutations correspond to deductions and it could be observed that even though mutations can affect the level of biological activity, their function is kept [5]. In the human phenylalanine hydroxylase (PAH) gene the most frequent mutations correspond to deductions too. The majority of these mutations result in deficient enzyme activity and cause hyperphenylalaninemia [7].

So, we have experimental evidence and we can say that in the B(X) algebra the deductions have a physico-chemical and biological meaning.

2.2 The Hamming Distance

The Hamming distance between two codons in the Hasse diagram reflects the difference between the physico-chemical properties of the corresponding amino acids. In Table 2 the average of the Hamming distance between the codon sets XAZ, XUZ, XCZ and XGZ is shown (X, Z $\in$ {A, C, G, U}).

The maximum distance corresponds to the transversions in the second base of codons. It is well known that such transversions are the most dangerous since they frequently alter the hydrophobic properties and the biological functions of proteins. Particularly, between codons of hydrophilic and hydrophobic amino acids, there are larger values of the Hamming distance.

Table 2. The average of the Hamming distance between the codon sets *XAZ, XUZ, XCZ* and *XGZ*

	XGZ	*XUZ*	*XAZ*	*XCZ*
XGZ	3	3	3	4
XUZ	3	2	4	3
XAZ	3	4	2	3
XCZ	4	3	3	2

The results of these assumptions have been verified in the experimental data [6]. Particularly, almost all reported mutations conferring drug resistance in HIV proteasa gene have a small Hamming distance with regard to the wild type, and similar results have been obtained with mutations in HIV reverse-transcriptase, the beta-globin gene and the PAH gene, named above.

Furthermore, we found that the small difference between the enzyme activities of the wild type and the mutant means a small Hamming distance between them. (Fig3). Generally, one can observe that as the Hamming distance between the wild type

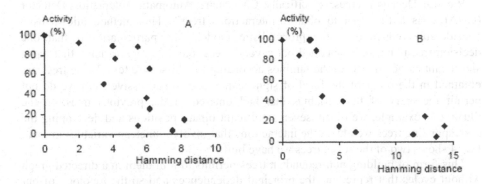

Fig. 3. Changes in the enzymatic activities of mutants versus the Hamming distance. *A:* Ability of HIV mutant proteases to process the Gag polyprotein [10]. *B:* DNA polymerase activity of the HIV reverse transcriptase [11]. Activity changes in the mutants are normalized with respect to the wild type enzyme.

increases, the mutant enzyme activity decreases. Such results are a consequence of the genetic code order. The Hamming distance between DNA bases is determined by their physico-chemical properties.

The arrangement of codons in the genetic code is such that the Hamming distance between codons are connected with the physico-chemical properties of amino acids. The experimental confrontations suggest that in the molecular evolution process, the mutation pathway tends to have the minimal Hamming distance between the wild type and the mutant genes (proteins) in each mutation step. These results advance another idea that the Boolean lattice could allow us to model the gene mutation process.

3 Possibilities of Applications to Sequence Analysis

Now we summarize two different types of applications for bioinformatics sequence analyses. Results are obtained with the use of the recommended coding system from Boolean algebras and by using Decision Trees complemented with Bayesian Networks.

3.1 Analysis of Mutant Sequences with a Bayesian Network

In order to illustrate the methodology, we use a data base with sequences of HIV-1-protease mutants, each one with 99 codons. It is assumed that these sequences have been previously grouped in "classes" or "families" according to certain criterion, that could be for instance, resistance level to an antiviral (High, Medium, Low). In the example shown clustering techniques to obtain 3 classes are used. Classes are considered the "dependent variable" if we approach the task as supervised machine learning problem. As "independent variables" of "features" we used the elements of the sequence coded as a binary pair according to the primal Boolean algebra of four bases. So we have 99*6=594 variables.

We use Decision Trees, specifically Chi-square Automatic Interaction Detector (CHAID) as a first tool to obtain "interactions trees". This method allows us a considerable reduction of the probabilistic model. The particularity is that one decision tree is not only obtained but several trees, one of every variable that has a significant association with the families according to Chi-square test. These trees are obtained in the order of the level of significance and in successive trees we do not permit the entry of the variables that had entered in the previous trees. In the illustrative example, we obtain seven significant binary positions and developing the corresponding trees we obtain the interactions that involve nineteen variables in total. Fig. 4 shows two of the seven trees we have built.

This form of building non-redundant trees permits to join them in a directed graph without cycles that represents the principal dependences and so the topology of our Bayesian network (Fig. 5). Now the conjoint distribution of *family* - as a function of $v1, v2, v3,...v594$ - is reduced to calculus with $v546, v323, v544, v324, v543, v500,...$ $v422, v296$ (nodes of the network) and involves a minimal number of conditional

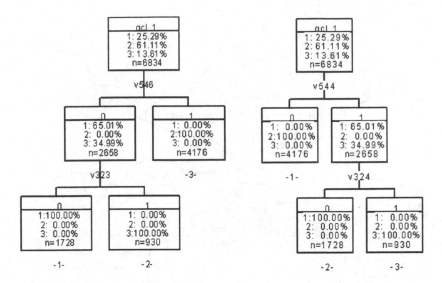

Fig. 4. On the left, the tree developed from the position *546* (*v546*) that interacts with position *323*. On the right, the tree developed from the position *544* interacting with the position *324*. The *percentages* in each node besides *1: 2: 3:* denote the proportion of mutant sequences in respective families 1 2 and 3. For instance, *100*% of the sequences with *v546*=0 and *v323*=0 belong to family 1. *100*% of the sequences with *v546*=0 and *v323*=1are in family 3 and *100*% of the sequence with *v546*=1 belong to family 2.

probabilities. They are easily obtainable with any statistical package. We used SPSS (Statistical Package for the Social Sciences)

For evidence propagation in the Bayesian network, a software ("ByShell", [12]) was elaborated. This software implements a propagation algorithm for multi-connected networks specifically a clustering tree algorithm that it is considered one of the less complexity exact propagation algorithms. Now new proposal in distributed form for this algorithm and approximated algorithms with simulation techniques are being studied [13], [14], [15], [16].

With the network propagation in ByShell we can predict the family for a new mutation or reciprocally, we can predict the content of certain positions in the DNA mutant sequence if we know previously the family and/or some other positions. For instance, if we have identified a sequence as belonging to the family 2, and we would know the position 296 corresponding to the second binary of the first base of codon 50th (296=49*6+2), we can arrive to the conclusion that this nucleotide base should be an A or a C because the algorithm answers that this position is 1 with probability 97%. If we also analyze the first binary number of this nucleotide (v295) we definitively conclude that this base is probably A, because v295 is always equal to 0 in these conditions.

The main two advantages of Bayesian network as supervised machine learning approach are precisely that the concepts of "dependent" and "independent" variables can be interchanged, and also the non-necessity of all information about the rest of the variables to predict one other. We can always obtain a prediction with certain

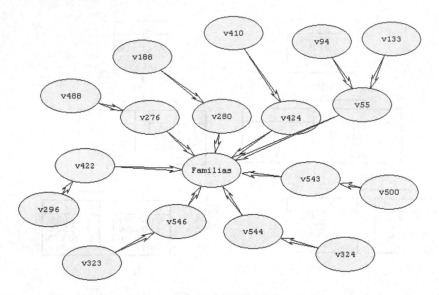

Fig. 5. Bayesian network topology, obtained from the join of seven non-redundant decision trees. One can see in particular the schematic representation of the *family* dependence on *v546* interacting with *v323*; or the *family* dependence on *v544* interacting with *v324*.

probability. It is clear the importance of these facilities in bioinformatics sequence analyses.

If we are studying of genes (proteins) mutations as in this example, these techniques can obviously be completed with the information about operators, deductions and Hamming distance in the Boolean algebra as we have shown with HIV protease in [17], or HIV reverse-transcriptase in [18]. There are interesting questions as the followings that can be answered. Suppose that you have a new sequenced mutation Is this a "theoretical mutation" in the sense of deductions in Boolean algebra? What are the next most probably mutations from this one? Is this mutation far (by using the Hamming distance point of view) from the wild type? Which are the minimum changes in the sequence we can expect for a draft change in their antiviral resistance, for instance?

However the next example can illustrate that this "simple codifying system" from Boolean algebra of 4 bases can be useful in the analyses of other types of interesting bioinformatics sequences, not necessarily proteins.

3.2 Classifying "True vs. False" Splice Sites in Human Sequences.

In this problem we use two data base sequences of nucleotide bases from the human genome and pretend to classify true vs. false splice sites: donors identified for the nucleotide pair "GT" and acceptors by the pair "AG". The first data base has sequences with the di-mers GT and contains known 6000 false and 1000 true donors. The other data base has sequences with the di-mers AG and contains also 6000 false and 1000 true cases of acceptors. Of course the length of the sequences of pairs "intron-exon" ("exon-intron") is very variable. So we began doing a statistical analysis about the distribution of exon (intron) length in order to fix a representative window with all the sequences of the same length. We obtained that it could be enough to take amounts of 80 nucleotide bases in the intron and 22 bases in the exon.

In both cases, we codify the positions before the queried di-mer as the primal Boolean algebra recommended, in such a way that we had 102 positions (splice included) and so 204 binary predictive variables. We had also the dependent variable for supervised learning (1: Positive splice, 0: Negative splice).

We used Decision Trees, in this case with the QUEST method implemented in SPSS, with 8 level of depth. After a 10 – fold cross validation, we obtained a final tree with all the data. Results are briefly commented.

Donors. In this case, we fixed 22 positions on the left and 78 positions after the pair GT in such a way that positions 23th and 24th with the pair GT are represented for instance for v23_1=0, v23_2=0 v24_1=1 v24_2=0. The Decision tree obtained has 5 terminal nodes in which the response is "Positive" (the rest is negative) and it is possible to prune this tree until 4 positive nodes without lose information. The most interesting thing as the results only depend on 6 nucleotide positions: two at left and four on the right. As shown in Table 3 the rules that lead to a positive prediction are mutually exclusive and characterize the donor patterns.

Table 3. Patterns around the pair "GT" corresponding to true donors

Patterns	-2	-1		+1	+2	+3	+4
1	A or C	G		A		T or C	
2		G	GT	A		A	
3		T or A or C		G or A	A or C	G	G or T
4		G		G or A	A or C	G	

Several statistical measures of predicting performance were calculated in classical form from TP (True Positives), TN (TN Negative), FP (False Positives) and FN (False Negatives). Particularly, we obtained FP%= 4.3%, FN%= 26.9%, accuracy: Ac=(TP+TN)/(TP+TN+FP+FN) = 92.4%, sensitivity: Se=TP/(TP+FN) = 73.1% and Specificity has been calculated in accordance with two criterions: Sp=TN/(TN+FP) = 95.6% and Sp'=TN/(TN+FN) = 95.5%

Acceptors. In this case, we fixed 20 positions on the left and 80 positions after the pair AG. The Decision tree obtained has 5 terminal nodes in which the response is "Positive" (the rest is negative). Patterns are shown in Table 4. The measures of performance classification were in this case as follow: FP%=5.1% FN%=36.2% Se=63.7% Sp=94.8% Sp'=94% Ac=90.4%

In general, these classifiers are not better than other reported in the literature [19], but they are good enough, specially to predict positive cases, and they are simpler because involve only one feature that is expressed as a Boolean expression in disjunctive normal form. It is clear also that we can obtain other decision trees and join them in a Bayesian neural network as explained above.

Table 4. Patterns around the pair "AG" corresponding to true acceptors

Patterns	-13	-12	-11	-10	-9	-8	-7	-6	-5	-4	-3	-2	-1	+1 a+18	+19
1	ToC							ToC	GoA	ToC	ToC		ToC	AG	
2		ToC	ToC				ToC		ToC	Go A	ToC		ToC		
3				ToC	GoA				ToC	ToC	ToC		ToC		
4					ToC				ToC	ToC	ToC		ToC		
5			ToC	ToC					ToC	ToC	GoA		ToC		ToC

4 Conclusions

Boolean algebraic structures of the sets of four nucleotide bases and codons are simple mathematical models in which Boolean operators, deductions, and Hamming distance have a physico-chemical and biological sense. It can help us to understand, the evolution process. It has been shown how to use the codifying system recommended from the Boolean algebras to solve some typical problems of bioinformatics sequence analyses for knowledge discovering by using Decision trees and Bayesian networks.

Acknowledgments. This work was developed in the framework of a collaboration program supported by VLIR (Flemish Interuniversity Council, Belgium). Particularly thanks are due to Yves Van de Peer and Yvan Saeys, Ghent University, by their help.

References

1. Crick, F. H. C.: The Origin of the Genetic Code. J. Mol. Biol. 1968, Vol. 38, No.3, (367-79).
2. Freeland, S., Hurst, L.: The Genetic Code is One in a Million. J. Mol. Evol. Vol. 47, No.3, (1998),.238-248.
3. Alf-Steinberger, C.: The Genetic Code and Error Transmission. Proc. Natl. Acad. Sci. USA, Vol. 64, No.2, (1969), 584-591.
4. Swanson, R.: A Unifying Concept for the Amino Acid Code. Bull. Math. Biol. Vol.46, No.2:, (1984), 187-203.
5. Sanchez, R., Grau R., Morgado, E.: A Genetic Code Boolean Structure I. Meaning of Boolean Deductions, Bull. Math. Biol. Vol. 67, (2005), 1-14
6. Sanchez, R., Grau, R., Morgado, E.: The Genetic Code Boolean Lattice, MATCH Commun. Math. Comput. Chem., 52, (2004), 29-46
7. Sánchez, R., Grau, R., Morgado E.: Genetic Code Boolean Algebras, WSEAS Transactions on Biology and Biomedicine, 1 (2004), 190-197
8. Friedman, S.M., Weinstein, I.B.: Lack of Fidelity in the Translation of Ribopolynucleotides. Proc. Natl. Acad. Sci., Vol. 52, (1964), 988996.
9. Parker, J.: Errors and Alternatives in Reading the Universal Ggenetic Code. Microbiol. Rev. Vol.53, No.3, (1989). 273-298.

10. Rose, R. E., Gong, Y., Greytok J. A., Bechtold, C. M., Terry, B. J., Robinson, B. S., Alam, M., Colonno, R. J., Lin, P.: Human Immunodeficiency Virus Type 1 Biral Background Plays a Major Role in Development of Resistance to Protease Inhibitors. Proc. Natl. Acad. Sci. Vol.93, No.4, (1996) 1648–1653.

11. Kim, B., Hathaway, T. R., Loeb, L. A.: Human Immunodeficiency Virus Reverse Transcriptase Functional Mutants Obtained by Random Mutagenesis Coupled with Genetic Selection in Escherichia Coli. J. Biol. Chem. Vol.271, No.9, (1996), 4872–4878.

12. Chávez, M. C., Rodríguez, L.O.: Bayshell, Software para crear redes Bayesianas e inferir evidencias en la misma, Registro de Software CENDA, 09358-9358, mayo, 2002, Published in http://uclv.edu.cu/Bioinformatics Group/Software

13. Castillo, E., Gutiérrez, J.M., Hadi, Ali S.: Expert Systems and Probabilistic Network Models, Springer-Verlag, New York, Inc. 1996.

14. Stuart, R., Norvig, P.: Inteligencia Artificial: Un Enfoque Moderno, Prentice Hall, México, 1996.

15. Williams, W. L., Wilson, R.C., Hancock, E.R.: Multiple Graph Matching with Bayesian Inference. Pattern Recognition Lett. Vol 38, (1998), 11-13

16. Hunter, L.: Planning to Learn About Protein Structure, in Hunter, L. (ed) Artificial Intelligence and Molecular Biology, AAAI Press Book, Cambridge (2003)

17. Grau, R., Galpert, D., Chavez, M. C., Sánchez, R., Casas, G., Morgado, E.,: Algunas Aplicaciones de la Estructura Booleana del Código Genético, Revista Cubana de Ciencias Informáticas, Vol 1 (2006), 16-30

18. Chavez, M. C., Casas, G., Grau, R., Sánchez, R: Statistical Learning Bayesian Networks from of Protein Database of Mutants, Proceedings of First International Workshop on Bioinformatics Cuba-Flanders 2006, Santa Clara, Cuba, (2006), ISBN: 959-250-239-0

19. Degroeve, S., Saeys, Y., De Baets, B., Rouzé, P., Van de Peer, Y.: Predicting Splice Sites from High-Dimensional Local Context Representations, Bioinformatics, 21-8 (2005), 1332-1338

Discovery of Gene Regulatory Networks in *Aspergillus fumigatus*

Reinhard Guthke, Olaf Kniemeyer, Daniela Albrecht,
Axel A. Brakhage, and Ulrich Möller

Leibniz Institute for Natural Product Research and Infection Biology – Hans Knoell Institute,
Beutenbergstr. 11a, 07745 Jena, Germany
{reinhard.guthke,olaf.kniemeyer,daniela.albrecht,
axel.brakhage,ulrich.moeller}@hki-jena.de
http://www.hki-jena.de

Abstract. *Aspergillus fumigatus* is the most important airborne fungal pathogen causing life-threatening infections in immunosuppressed patients. During the infection process, *A. fumigatus* has to cope with a dramatic change of environmental conditions, such as temperature shifts. Recently, gene expression data monitoring the stress response to a temperature shift from 30 °C to 48 °C was published. In the present work, these data were analyzed by reverse engineering to discover gene regulatory mechanisms of temperature resistance of *A. fumigatus*. Time series data, i.e. expression profiles of 1926 differentially expressed genes, were clustered by fuzzy c-means. The number of clusters was optimized using a set of optimization criteria. From each cluster a representative gene was selected by text mining in the gene descriptions and evaluating gene ontology terms. The expression profiles of these genes were simulated by a differential equation system, whose structure and parameters were optimized minimizing both the number of non-vanishing parameters and the mean square error of model fit to the microarray data.

1 Introduction

Pathogenic fungi need to deal with a variety of environmental challenges, such as temperature shifts, during the course of an infection. The ability to meet these challenges requires the expression of many specific genes. *Aspergillus fumigatus* has become the most important airborne fungal pathogen of humans causing pneumonia and invasive disseminated disease with high mortality in the immunocompromised hosts. The lack of effective treatments results in a very high mortality rate of 30 % to 90 % [1]. The complete 29.4-megabase genome of the clinical isolate Af293 of *A. fumigatus*, which consists of eight chromosomes containing 9,926 predicted genes, has been sequenced recently [2]. The thermotolerance of *A. fumigatus* is a trait critical to its ability to thrive in the human body. To investigate the adaptation of this fungus to higher temperatures, gene expression was examined throughout a time course upon shift of growth temperatures from 30 °C to 37 °C and 48 °C (representing temperatures in the human body and compost, respectively) [2]. The time series of

K. Tuyls et al. (Eds.): KDECB 2006, LNBI 4366, pp. 22–41, 2007.

1926 differentially expressed genes were grouped into 10 clusters using k-means clustering method (Supplementary Table S2 of [2]).

Here we optimized the clustering and interpreted the expression profiles of representative genes by an optimized dynamic linear network model on the background of gene regulation. Different dynamic models have been constructed using four distinct sets of genes found as representatives of the clusters by the application of several selection criteria.

2 Material and Methods

2.1 Data

Gene expression profiles monitoring the stress response of *A. fumigatus* to a temperature shift were obtained from [2]. The raw data measured using the *A. fumigatus* Af293 DNA amplicon microarray containing 9,516 genes are available from the Internet (http://www.ebi.ac.uk/arrayexpress, with accession numbers A-MEXP-205, E-MEXP-332 and E-MEXP-333). We used the pre-processed data representing logarithmized ratios (log-ratios) of the expression intensities of 1926 genes at six time points t (= 0, 15, 30, 60, 120, 180 min) before and after the temperature shift (Table S2 of [2]). We focused the study on the temperature shift from 30 °C to 48 °C because the mean absolute log-ratios are significantly higher after this shift than after the shift to 37 °C. Different methods for the imputation of missing data were applied, including the Gaussian mixture clustering method (*gmc*) [3], the k-nearest neighbor algorithm (*knn*) [4], and the tri-level alternating optimization method (*tao*) [5, 6]. The log-ratios at $t = 0$ (before the temperature shift) were subtracted from the respective time series data, i.e. only differences with respect to the pre-perturbation state were considered.

2.2 Clustering and Cluster Validation

The time series data were scaled between their respective absolute temporal extreme values to focus subsequent cluster analysis on the *qualitative* behavior of the expression profiles. The fuzzy C-means (*fcm*) algorithm [7] was used for clustering (number of clusters $C = 2, ..., 15$; fuzzy exponent 1.2; maximum number of iterations = 300; minimum cost function improvement = 10^{-8}). The optimum number of clusters C was estimated by the vote of 36 cluster validity indices (CVIs): 18 generalizations of Dunn's index [8] were computed and the same 18 generalizations were applied to the Davis-Bouldin index [9]. These indices capture different aspects of a clustering structure. In order to reduce a bias of the biological sample, C was alternatively estimated by a novel method based on maximum partition stability under a nearest-neighbor (NN) resampling [10].

2.3 Selection of Cluster-Representative Genes by Gene Description Text Mining

For each cluster ($c = 1,..., C$) one representative gene was selected evaluating the description string S_i and the maximum absolute value of log-ratio assigned to the respective gene i. The following selection criteria were used:

- The representative gene is annotated with known physiological function containing the motif s_c, that was found within the gene description strings S_i for many genes ($m_c \geq 5$) of the respective cluster c and for no or only few genes m_j belonging to the other clusters j ($m_j < 3$, $j \neq c$); genes were sorted within a cluster c by the score $M_c = m_c - \Sigma_{j \neq c}\, m_j$; the string (motif) s_c with the highest score M_c was selected.
- The representative gene shows an expression profile with at most one missing value.
- The representative gene is characterized by a high temporal maximum of its absolute log-ratio value.

Subsequently, the expression profiles of the selected C genes were used for dynamic modeling.

2.4 Selection of Cluster-Representative Genes Using GO Terms

As an alternative to the analysis of the description string S_i assigned to gene i, we analyzed sets of GO terms (Gene Ontology, [11]) assigned to the genes i of *A. fumigatus*. These sets of GO terms were formed by the set of matching GO terms A found in the CADRE database [12] for gene i as well as the sets of GO terms B that subsume the GO terms A. For each cluster c and each GO term T we calculated the number m_c of genes, which belong to the cluster c and were assigned to GO term T. The GO terms were sorted within each cluster c according to the score $M_c = m_c - \Sigma_{j \neq c}\, m_j$; The GO term T_c with the highest score M_c was selected. The cluster-representative gene

- belongs to cluster c,
- is assigned to the selected GO term T_c ,
- shows an expression profile with at most one missing value,
- is characterized by a high temporal maximum of its absolute log-ratio value.

2.5 Dynamic Modeling Using a Search Strategy

The dynamics of hypothetic gene regulatory networks was modeled by a system of linear differential equations. The general mathematical form reads

$$\frac{dx_i(t)}{dt} = \sum_{j=1}^{C} w_{i,j} \cdot x_j(t) + b_i \cdot u(t) \tag{1}$$

where $x_i(t)$ is the expression (log-ratio) of representative gene i (= 1,...,C) at time t, w_{ij} denotes the gene-gene interaction matrix and b_i represents the perturbation response vector. $u(t)$ is the Heaviside step function: $u(t<0) = 0$ and $u(t \geq 0) = 1$ representing the temperature shift from 30 °C to 48 °C. The system is assumed to be at equilibrium prior to the perturbation, i.e. $dx_i(t<0)/dt = x_i(t<0) = 0$.

Genetic networks are known to be sparsely connected [13]. The aim of dynamic modeling and network reconstruction is thus to find a minimal set of relevant (i.e. non-zero) model parameters (w_{ij} and b_i) that are required to achieve an adequate fit to the expression data at hand minimizing the mean square error (*mse*). For this reverse engineering approach we used the network generation method recently described [14]. Here, the maximum dynamic order was set to one. The maximum allowed sub-model error was set to 0.01.

3 Results and Discussion

3.1 Clustering and Cluster Validation

A small number of 221 values (1.9 % of the 5*1926 values) of the data set is missing. The method applied to impute these missing values influences the results of clustering in detail only, but it does not influence the main conclusions. In particular, in each case we estimated four clusters. The majority of generalizations of Dunn's index and of the Davis-Bouldin index voted for two or four distinct clusters as shown in Figure 1a after *gmc* imputation. The results of the nearest-neighbor resampling robustly indicated highly stable partitions of up to four clusters (Figure 1b). Thus, the expression profiles were grouped into four clusters that we labeled by '*Increasing*', '*Minimum*', '*Maximum*' and '*Decreasing*' (Figure 2 and Table 1).

After *gmc* imputation and *fcm* clustering, 247 (N_c in Table 1) genes belong to cluster one labeled by '*Increasing*', whereas 336 (N_c' in Table 1) genes do so after *tao* imputing and 244 genes are the same in both cases. Applying *gmc* imputation before clustering, 443 genes are assigned to cluster two (labeled by "Minimum") while the usage of *tao* imputation leads to the assignment of 433 genes of which 432 genes coincide with those selected after *gmc* imputation. For cluster three (labeled by "Maximum") the application of *gmc* imputation leads to the assignment of 480 genes while *tao* imputation entails the assignment of 423 genes of which 421 coincide with those obtained after *gmc* imputation. 59 genes were shifted from cluster three to cluster one when the *tao* imputation was applied instead of the *gmc* method. As for cluster four (labeled by "Decreasing") *gmc* imputation leads to the assignment of 427 genes, *tao* imputation to the assignment of 418 genes and a number of 416 genes coincide.

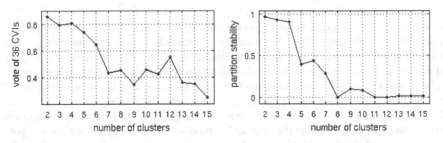

Fig. 1. a) *Left*: Weighted vote of the 36 cluster validity indices (CVIs). A vote of 1.0 for the *C*-cluster partition means that all indices had their global extremum at the value of *C*. A vote < 1 indicates that some CVIs exhibited a value different from their global extremum. b) *Right:* Partition stability under the nearest-neighbor (NN) resampling technique with 10 NNs. A value of 1.0 means that all clusters of 10 resample partitions showed at least an 80% overlap with a cluster of the other partitions, respectively. Values < 1 indicate lower cluster overlap. All results were obtained for the dataset completed by the *gmc* imputing method.

3.2   Selection of Cluster-Representative Genes by Gene Description Text Mining

In order to identify representative genes for each cluster c, we searched for strings s_c that were found as part of the description of many (m_c) of the N_c genes and of only few (m_c') genes of the other clusters (c' ≠ c). Maximizing the difference of both gene numbers ($M_c = m_c - m_c'$) for each cluster c, we found for three clusters ($c = 1, 2,$ and 3) the strings *'peroxi'*, *'ribosom'* and *'heat shock'*, denoting genes coding for

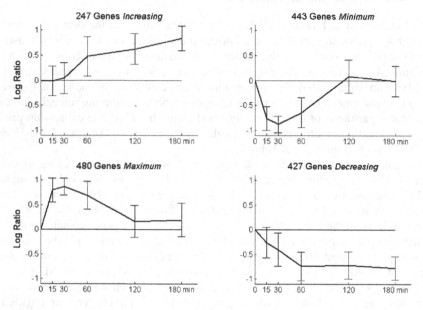

Fig. 2. Result of *fcm* clustering with four clusters after imputation of the missing values by the *gmc* method: mean scaled gene expression profiles with standard deviation averaged over the N_c genes for the respective cluster (Table 1)

Table 1. Number N_c of genes belonging to cluster c ($c = 1,...,C$) with the membership degree greater 50% after *gmc* imputation. The numbers N'_c were obtained after imputation by the *tao* method. The maximum value $mLRN_c$ was estimated over the absolute log-ratios of the N_c genes and the six time points t. The string motifs s_c denote representative gene functions (*'peroxi'* – oxidative stress, *'ribosom'* – ribosomal protein, *'heat shock'* – heat shock proteins, *'sterol'* – sterol biosynthesis). The string s_c is contained in the description S_i of m_c genes of the Cluster c and in the description of m_c' (= m_c -M_c) genes belonging to the other clusters. The maximum value $mLRm_c$ was estimated over the absolute log-ratios of the m_c genes and the six time points. Cluster means and standard deviations are shown in Figure 2.

c	N_c	N'_c	Label	$mLRN_c$	s_c	m_c	m_c'	M_c	$mLRm_c$
1	247	336	*Increasing*	4.94	*'peroxi'*	9	0	9	2.01
2	443	433	*Minimum*	3.06	*'ribosom'*	23	5	18	3.06
3	480	423	*Maximum*	6.94	*'heat shock'*	7	2	5	6.94
4	427	418	*Decreasing*	4.69	*'sterol'*	7	2	5	2.86

oxidative stress proteins, ribosomal proteins and heat shock proteins, respectively (see Table 1 and 2). These gene functions are reasonable, since the respective genes are involved in temperature shift response or are related to a general stress resistance. Figure 3 shows the scaled expression profiles of those 9, 23 and 7 genes (see Table 2 for $c = 1$, 2 and 3), which belong to clusters one, two and three and code for oxidative stress proteins, ribosomal proteins and heat shock proteins, respectively.

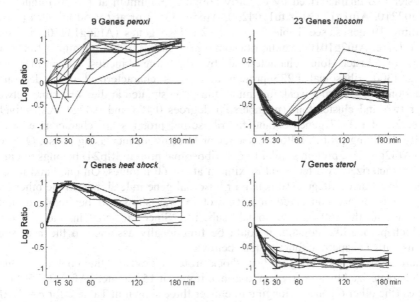

Fig. 3. Scaled expression profiles of the genes belonging to the clusters c and coding for oxidative stress proteins ('*peroxi*', $c = 1$), ribosomal proteins ('*ribosom*', $c - 2$), heat shock proteins ('*heat shock*', $c = 3$) and sterol biosynthesis enzymes ('*sterol*', $c = 4$) listed in Table 2. The mean kinetics are shown as thick lines.

Among the 1926 differentially expressed genes, there are 10 genes annotated by descriptions containing the string '*peroxi*'. Nine of them belong to cluster one characterized by increasing kinetics (see Table 2 and Figure 3). The remaining gene coding for MnSOD (Afu1g14550, manganese superoxide dismutase) belongs to none of the four clusters because it is at the borderline between cluster one and cluster three (i.e., the respective membership degrees are 0.448 and 0.451, which do not suffice to surpass the required threshold 0.5), due to the late weak declination of its expression profile. After *tao* imputation all 10 genes that are annotated with descriptions containing the string '*peroxi*' are members of cluster one (note that after *tao* imputation cluster one contains 89 genes more than after *gmc* imputation; see N_c and N_c' for $c = 1$ in Table 1). Thus, the expression profiles of all genes coding for oxidative stress proteins were increasing after the temperature shift.

We selected the catalase/peroxidase HPI (*cat2*) as the representative gene of the 'Increasing' cluster because it is implicated in *A. fumigatus*' pathogenicity (see [15] and Table S1 in [2]). The gene *cat2* is ranked at the second position in Table 2 for cluster one. The gene *kat1* ranked at the first position was not selected as the cluster representative gene because it codes for a protein that is not involved in the oxidative stress response; it is only located in peroxisomes.

Among the 1926 differentially expressed genes, there are 29 genes annotated with descriptions containing the string '*ribosom*'. Two of them are non-ribosomal (Afu5g10120, monomodular non-ribosomal peptide synthetase; Afu3g03350, nonribosomal peptide synthetase NRPS). 23 genes of the remaining 27 genes belong to cluster two characterized by an early temporal minimum at $t = 30$ minutes (for Afu1g02210, Afu1g12730, Afu1g04230, Afu5g08350) or at $t = 60$ minutes (for the remaining 19 genes, see Table 2 for $c = 2$). Two genes (Afu6g13250, ribosomal protein L31e; Afu2g10100, acidic ribosomal protein P2) of the remaining four genes belong to cluster four characterized by declining kinetics. A further gene (Afu3g12300, ribosomal L22e protein family) is characterized by a decreasing expression profile with a weak minimum, thus it is situated at the borderline between cluster two and cluster four (membership degrees 0.453 and 0.4724, respectively). Hence, 26 of the 27 genes coding for ribosomal proteins are characterized by an initially declining profile followed sooner or later by an increasing phase. Only one gene (Afu2g02710, protein similar to 60S ribosomal protein Rlp24) belongs to cluster three characterized by a temporal maximum at $t = 60$ minutes. On one hand it could be speculated that Afu2g02710 is not a ribosomal gene indeed, but on the other hand this gene is highly conserved and its protein was shown to be involved in the assembly of the 60 S ribosomal subunit. Since it contains a conserved metallochaperone-like domain, it could be functionally assigned to the heat shock proteins with a chaperone function (see below).

There are 10 genes coding for heat shock proteins. Seven of them belong to cluster three characterized by an early temporal maximum at 15 minutes (Afu2g11750) or 30 minutes (the other 6 genes belonging to cluster three shown in Table 2 for $c = 3$) after temperature shift. The expression profiles of two genes (Afu5g10270, heat shock protein, HSP20 family; Afu6g12450 12, kDa heat shock protein) are characterized by a late temporal maximum at 60 minutes after temperature shift. Thus, 9 of the 10 heat shock genes are characterized by an expression profile with a temporal maximum. The remaining one (Afu8g03930, heat shock protein 70) belongs to cluster two characterized by a temporal minimum.

For the clusters two and three, we selected as representatives those genes which code for the large subunit ribosomal protein L3 (*rpl3*, Afu2g11850) and the heat shock class I protein (*hsp30*, Afu6g06470), respectively, due to their maximum absolute log-ratio value mLR estimated as the greatest within the respective group c (Table 2). The ribosomal protein RPL3 is known as an *A. fumigatus* allergen (Table S4 in [2]). The up-regulation of the heat shock protein HSP30 was verified by proteome analysis, too (data not shown).

Among the 1926 differentially expressed genes, there are 9 genes annotated with descriptions containing the string '*sterol*'. The expression profiles of all of these genes are decreasing immediately after the temperature shift. Seven of them belong to cluster four (see Table 2 and Figure 3). The remaining two genes (Afu1g04720, C-8 sterol isomerase erg-1; Afu1g07140, c-24 (28) sterol reductase) belong to cluster two whose gene expression profiles initially declines too, but after passing a minimum (at $t = 60$ min) the profiles increase again.

Table 2. Genes belonging to the clusters c annotated with a gene description string S_i containing the pattern s_c (see Table 1). Descriptions of genes selected as cluster representatives are underlined. *mLR* - maximum absolute value of log-ratios.

c	Gene ID	Gene Description String S_i	mLR
1	Afu1g12650	3-ketoacyl-CoA thiolase peroxisomal precursor (*kat1*)	2.01
	Afu8g01670	Catalase/peroxidase HPI (*cat2*)	1.95
	Afu4g09110	Cytochrome c peroxidase precursor	1.50
	Afu8g05160	Peroxisomal membrane protein pex13 (peroxin-13)	1.38
	Afu4g11580	Mn-superoxide dismutase	1.37
	Afu6g04040	Peroxisomal D3,D2-enoyl-CoA isomerase	1.17
	Afu1g13840	22 kDa peroxisomal membrane protein	1.09
	Afu7g06100	Acyl-coenzyme A oxidase I, peroxisomal, component A	1.08
	Afu5g04310	Peroxisomal membrane protein (pmp47)	0.86
2	Afu2g11850	Large subunit ribosomal protein L3 (*rpl3*)	3.06
	Afu6g03830	Ribosomal protein L14	2.72
	Afu1g10510	Ribosomal protein L35-like protein	2.66
	Afu3g06960	60S ribosomal protein l21	2.64
	Afu2g02150	Ribosomal protein S10	2.57
	Afu3g13320	Ribosome-associated protein	2.54
	Afu6g13550	Ribosomal protein S13p/S18e	2.50
	Afu3g06970	Ribosomal protein S9 (S7)	2.49
	Afu6g12660	40S ribosomal protein	2.46
	Afu1g04530	Ribosomal L18ae protein family	2.42
	Afu2g08130	Ribosomal protein L41	2.35
	Afu2g10440	Ribosomal protein S14.e	2.35
	Afu3g08460	60S ribosomal protein l37b	2.33
	Afu2g10090	Ribosomal protein S15 (S12)	2.28
	Afu6g11260	Ribosomal protein L26	2.14
	Afu1g11710	60S ribosomal protein L1	2.14
	Afu6g12720	Probable ribosomal protein S29. cytosolic	1.95
	Afu1g02210	Required for biogenesis of the 60S ribosomal subunit	1.73
	Afu7g05290	Ribosomal protein S13, cytosolic	1.65
	Afu1g09440	Ribosomal protein S23 (S12)	1.61
	Afu1g12730	Mitochondrial 60S ribosomal protein l3 precursor	1.09
	Afu1g04230	Mitoribosomal protein YmL27	0.88
	Afu5g08350	Ribosomal protein S16	0.36
3	Afu6g06470	Heat shock protein, class I (*hsp30*)	6.94
	Afu1g07440	Heat shock protein 70	3.91
	Afu1g15270	Heat shock protein CLPA	3.39
	Afu3g14540	30 kDa heat shock protein	2.96
	Afu5g04170	Heat shock protein 80	2.90
	Afu1g12610	Heat shock protein Hsp88	2.85
	Afu2g11750	Heat shock protein	2.34
4	Afu4g06890	Cytochrome P450 sterol 14-alpha-demethylase (*erg11*)	2.86
	Afu5g07780	Squalene epoxidase; ergosterol biosynthesis	2.78
	Afu7g03740	14-alpha sterol demethylase	2.76
	Afu1g03150	C-14 sterol reductase	2.59
	Afu2g00320	Sterol delta 5,6-desaturase ERG3	2.45
	Afu8g02440	C-4 methyl sterol oxidase	2.25
	Afu1g05720	C-14 sterol reductase	1.30

The string *'reverse transcriptase'* is contained within the descriptions of six genes belonging to cluster four and was not found in the description of genes belonging to any of the other clusters. The resulting score $M_4 = 6$ for the string *'reverse transcriptase'* is greater than the score $M_4 = 5$ for *'sterol'*. However, the relevance of this gene function to the temperature shift response of the fungus has to be investigated and discussed by further work. Here, we selected the string *'sterol'* that was ranked at the second position of motifs typical for cluster four. The formation of ergosterol is essential for fungal growth and vital for fungal cell membrane integrity. The gene *erg11* (Afu4g06890) coding for the cytochrome P450 sterol-14-alpha-demethylase, which is the target for azole antifungal drugs [16], was found in Table 2 with a maximum absolute log-ratio *mLR*. This gene was selected as representative for cluster four.

3.3 Selection of Cluster-Representative Genes Using GO Terms

Utilizing the CADRE database [12], we were able to find sets of matching GO terms for 1907 of the 1926 differentially expressed genes. Among those sets, member numbers varied between zero and thirteen. The remaining 19 genes could not be found in the CADRE database, therefore we assumed that no GO terms were available. After identifying superclasses in the matching GO, herein after referred to as *GO superclasses*, the number of those plus the number of previously found matching GO terms varied between zero (for 898 genes) and 31 (for two genes) as shown in Figure 4. For 137 genes a number of 7 GO terms were assigned. Altogether, 1849 GO terms were assigned to one or more of the 1926 genes. Taking into consideration that also subsets of GO terms may have a common GO superclass, the number of maximally 31 associated GO terms is reasonable.

Table 3. Number N_c of genes belonging to cluster c ($c = 1,...,C$) with membership degree greater 50% after *gmc* imputation (as in Table 1). T_c is the GO term (GO-name and GO-id) with score $M_c = m_c - m_c'$ maximized for the respective cluster c. As for score M_c, m_c refers to the frequency of assignment of GO term T_c to members (genes) of cluster c while m_c' is the number of assignments of the respective GO term to members of clusters other than c.

c	N_c	Label	T_c		m_c	m_c'	M_c
1	247	*Increasing*	*Peroxisomal part*	GO:0044439	9	5	4
2	443	*Minimum*	*Nuclear part*	GO:0044428	51	17	34
3	480	*Maximum*	*Protein folding*	GO:0006457	14	7	7
4	427	*Decreasing*	*Glucan metabolism*	GO:0006073	6	0	6

Table 3 shows those representative GO terms T_c to which a high number m_c of genes belonging to cluster c is assigned and to which only a low number m_c' of genes belonging to the other clusters is assigned. In comparison to the results mentioned in chapter 3.2 obtained by text mining, here we found the GO term *'peroxisomal part'* instead of the string motif *'peroxi'* for cluster one, the GO term *'nuclear part'* instead of the string motif *'ribosom'*, the GO term *'protein folding'* instead of the string motif *'heat shock'*, and the GO term *'glucan metabolism'* instead of the string motif *'sterol'*.

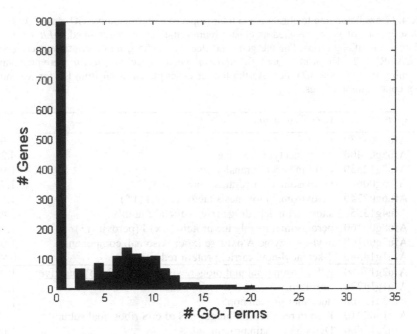

Fig. 4. Distribution ('histogram') of the number of GO terms (i.e., matching GO and GO superclasses) for the 1926 differentially expressed genes

These GO terms and the respective motifs are related physiologically to each other: Oxidase enzymes involved in the removal of reactive oxygen are located in peroxisomes; ribosomes are assembled in the nucleus; principle heat shock proteins are chaperones that assist in protein folding; and glucans form part of the fungal cell wall.

Table 4 shows the m_c genes (m_c =9, 51, 14 and 6 for the four clusters, respectively; see Table 3), sorted according to the maximum absolute value of log-ratios (mLR). The genes with the highest mLR were selected as representatives, i.e. *fadD35* for cluster one, *nip7p* for cluster two, *clpB* for cluster three and *mutA* for cluster four.

For cluster two, we not only studied the best matching GO term *'nuclear part'* (GO:0044428) but also the four terms that follow in the ranking according to score M_c as shown in Table 5. A common set of 23 genes is assigned to all of the first four GO terms shown in Table 5. This set is characterized by the process *'rRNA processing'* and the cellular component *'Nucleolus'* (the other two GO terms are superclasses of the latter term). The nucleolus, which is a substructure within the nucleus, is the site of rRNA processing and ribosome assembly. Thus, we can conclude that the expression of genes involved in rRNA processing and ribosome assembly follows mostly the time course of cluster two, characterized by a temporal minimum. Among the 23 genes that belong to cluster two and are involved in the nucleolar rRNA processing the gene *nip7p* is the one specified by the highest maximum log-ratio value mLR (= 1.99). Hence, this gene was selected as representative for cluster two.

Using the so called 'specificity' ($N_c = m_c /(m_c + m_c')$) as an alternative score, the fifth GO term in Table 5 *'RNA binding'* was found to be the most representative GO

Table 4. Genes belonging to cluster c and their respective representative GO term T_c (see Table 3). Descriptions of genes selected as cluster representatives are underlined. mLR - maximum absolute value of log-ratios. For the genes labeled by *) the gene descriptions were updated from CADRE [12]. The strings '*peroxi*', '*ribosom*', '*heat shock*' or '*glucan*' were not found by text mining (see chapter 3.2), because the former descriptions taken from [2] for text mining did not contain these strings.

c	Gene ID	Gene Description	mLR
1	Afu5g08470	<u>fadD35</u>	4.94
	Afu5g07400	phenylacetyl-CoA ligase	4.20
	Afu2g12530	carnitine acetyl transferase	1.96
	Afu5g00640	peroxisomal dehydratase, putative *)	1.76
	Afu6g07740	peroxisomal biogenesis factor (PEX11) *)	1.55
	Afu4g13550	short chain dehydrogenase/reductase family	1.40
	Afu8g05160	peroxisomal membrane protein pex13 (peroxin-13)	1.38
	Afu7g06100	acyl-coenzyme A oxidase I, peroxisomal, component A	1.08
	Afu1g14380	3-ketoacyl-acyl carrier protein reductase	0.98
2	Afu2g17060	<u>60S ribosome subunit biogenesis protein (Nip7p), putative</u>	1.99
	Afu1g14220	fibrillarin	1.81
	Afu3g13400	nucleolar protein nop5	1.78
	Afu1g02210	Protein required for biogenesis of 60S ribosomal subunit *)	1.73
	Afu2g12880	DUF663 domain protein	1.66
	Afu2g16260	putative microtubule-associated protein	1.64
	Afu3g09600	sik1 protein	1.51
	Afu5g13050	kinesin	1.50
	Afu1g12000	nuclear and cytoplasmic polyadenylated RNA-binding protein pub1	1.48
	Afu5g12100	pmt2 methyltransferase	1.47
	...	... (further 41 genes, not shown)	
3	Afu1g11180	<u>ATP-dependent Clp protease, ATP-binding subunit ClpB</u>	4.91
	Afu7g01860	activator of Hsp70 and Hsp90 chaperones	3.92
	Afu1g07440	heat shock protein 70	3.91
	Afu4g11330	Aha1 domain family	3.74
	Afu2g02050	peptidyl-prolyl cis-trans isomerase	3.52
	Afu2g09290	mitochondrial protein HSP60, putative *)	3.16
	Afu5g04170	heat shock protein 80	2.90
	Afu1g12610	heat shock protein Hsp88	2.85
	Afu5g13920	p21 protein	2.56
	Afu6g10700	Hsp10	2.47
	Afu2g11750	heat shock protein	2.34
	Afu2g02700	mitochondrial DnaJ chaperone (Tim14), putative	1.51
	Afu3g00990	flavin-binding monooxygenase, putative (T3P18.10)	1.48
	Afu6g10480	F22M8.7 protein	1.37
4	Afu8g06360	<u>mutanase (MatA, alpha-1,3-glucanase) *)</u>	2.75
	Afu1g16190	allergen (glucanase Crf1) *)	2.45
	Afu1g03350	mutanase (alpha-1,3-glucanase) *)	2.03
	Afu7g08510	alpha-1,3-glucanase *)	1.61
	Afu6g08510	Crh-like protein (glucanase) *)	1.15
	Afu5g10540	1,4-alpha-glucan branching enzyme	1.05

term for cluster two with 29 true positives and only 3 false negative results. Seven genes, shown in Table 6, are assigned to all of the five GO terms. Hence, the gene coding for the *rRNA processing protein Bystin* could be used as alternative to *nip7p* as representative of cluster two. However, due to the relative low maximum log-ratio of 1.46, that option will not be made use of in the following analyzes.

Table 5. GO terms T_c assigned to genes of cluster two ($c = 2$) with high score M_c; alternative score $N_c = m_c /(m_c + m_c')$; the other symbols as in Table 3

T_c		m_c	m_c'	M_c	N_c
Nuclear part	GO:0044428	51	17	34	0.75
Intracellular non-membrane-bound organelle	GO:0043232	43	10	33	0.81
rRNA processing	GO:0006364	41	10	31	0.80
Nucleolus	GO:0005730	36	9	27	0.80
RNA binding	GO:0003723	29	3	26	0.91

Table 6. Genes belonging to cluster two and to the five most representative GO terms T_c shown in Table 5. The gene descriptions were updated using CADRE [12].

Gene ID	Gene Description	mLR
Afu3g04110	rRNA processing protein (Bystin)	1.46
Afu3g05490	Nrap protein superfamily (nucleolar RNA-associated	1.35
Afu4g13690	protein)	1.31
Afu7g03690	small nucleolar ribonucleoprotein snoRNP protein (gar1)	0.99
Afu8g04790	G-patch RNA maturation protein (Gno1), putative	0.98
Afu2g16040	ribosome biogenesis protein, putative	0.81
Afu2g05930	rRNA biogenesis protein, putative	0.57
	small nucleolar ribonucleoprotein complex subunit, putative	

3.4 Dynamic Modeling

Model A
The measured expression profiles of the selected genes *cat2*, *rpl3*, *hsp30* and *erg11* were simulated by the differential equation system (2) with the variables x_1, x_2, x_3 and x_4, respectively. Figure 5 visualizes the network structure of the model. The results of model simulation are shown in Figure 6.

$$\frac{dx_1}{dt} = -0.0268 \cdot x_1 + 0.0431 \cdot u(t) \tag{2}$$

$$\frac{dx_2}{dt} = -0.0618 \cdot x_4 - 0.1580 \cdot u(t)$$

$$\frac{dx_3}{dt} = 0.2315 \cdot x_4 - 0.0245 \cdot x_3 + 0.7269 \cdot u(t)$$

$$\frac{dx_4}{dt} = -0.0599 \cdot x_1 - 0.1658 \cdot u(t) \cdot \cdot$$

Randomly disturbed input data were used for a model validation analysis by perturbation to assess the impact of measurement errors and to test the reliability of the structures generated. The analyzes were repeated 10^4 times using input data obtained by adding normal distributed random deviates with a standard deviation σ. With σ = 0.1 the structure with six links shown in Figure 4 was confirmed in 43 % of the cases. Five links without the activation link from the perturbation by temperature shift to *cat2* were confirmed in 82 % of all cases. Four links without the link from temperature to *cat2* and from *erg11* to *rpl3* were found in 90 % of the randomly repeated calculations (59 % for σ = 0.5). These four links from temperature to *hsp30*, *erg11* and *rpl3* and from *erg11* to *hsp3* are highlighted in Figure 4 by thick lines. In all 10^4 cases only the links from temperature to *hsp30* (activation) and *erg11* (inhibition) were found.

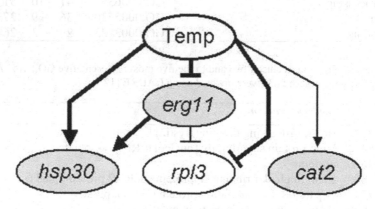

Fig. 5. Structure of the dynamic model A (2) identified by the model fit shown in Figure 6 (*Temp*: Temperature shift; *erg11*: cytochrome P450 sterol 14-alpha-demethylase, enzyme involved in ergosterol biosynthesis; *cat2*: catalase/peroxidase; *rpl3*: large subunit ribosomal protein L3; *hsp30*: heat shock protein class I). The arrows represent stimuli or activations. The T-shaped links (⊥) represent inhibitions. Grey boxes denote elements with negative $w_{i,j}$ (decay or self-regulation). The thick links indicate the connections confirmed by resampling.

Model B

The regulation of *cat2* was reconstructed vaguely due to the erratic data of the respective expression profile, i.e. the low value at *t* = 30 min and high value at *t* = 60 min are questionable (see Figure 6). Therefore, we selected alternatively for cluster one gene *kat1* as representative, which is the gene with the highest maximum logratio among the nine genes labeled by '*peroxi*' that belong to cluster 1 (the first gene shown in Table 2). Figures 7 and 8 show the reconstructed alternative network and the model fit results. The four confirmed links of the structure of model A (thick lines in Figure 5) were also found in the structure of model B (Figure 7). The structure shown in Figure 7 was confirmed in 84 % of the network reconstruction calculations repeated with randomly disturbed input data (σ = 0.1). Simulating model B with *N* = 11 parameters, the fit error was *mse* = 1.7, whereas model A with *N* = 9 parameters

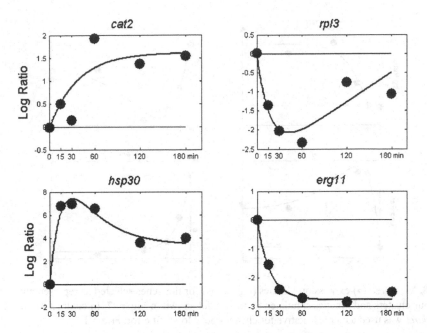

Fig. 6. Measured (•) expression kinetics (log-ratios) for the genes selected as representatives of the four clusters ($c = 1, 2, 3,$ and 4; Table 1) and kinetics simulated by the differential equation system (2), whose network structure is shown in Figure 5. Model fit error $mse = 2.9$.

leads to a fit error of $mse = 2.9$. According to the χ^2-criterion ($\chi^2 = mse / (n - N)$) and with the number $n = 24$ of measured data (i.e., four variables measured at six time points) model B is superior to model A. The model fit could not be improved using a higher dynamic order.

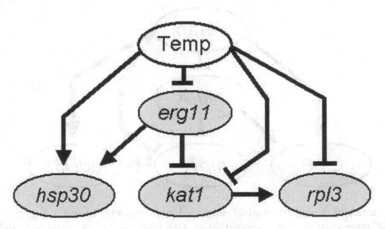

Fig. 7. Structure of the dynamic model B identified by the model fit shown in Figure 8: In opposition to Figure 5 gene *kat1* (3-ketoacyl-CoA ketothiolase / peroxisomal precursor) was selected as the representative of gene cluster one (labeled by '*Increasing*' and '*peroxi*') instead of *cat2* (catalase/peroxidase). Symbols as in Figure 5.

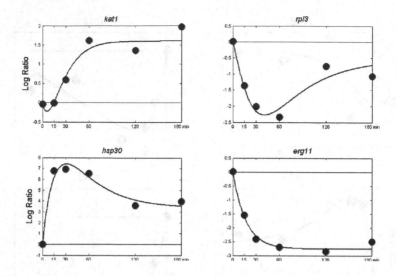

Fig. 8. Measured (•) expression kinetics (log-ratios) for the genes selected as representatives of the four clusters and kinetics simulated by the model shown in Figure 7. In contrast to Figure 6 gene *kat1* was used as representative for cluster one. Model fit error: *mse* = 1.7.

Model C

In chapter 3.3, the genes *fadD35, nip7p, clpB* and *mutA* were selected by GO term analysis as representatives for the clusters assigned to the phenomenological labels *'Increasing'* / *'Peroxisomal part'*, *'Minimum'* / *'Nuclear part'*, *'Maximum'* / *'Protein folding'* and *'Decreasing'* / *'Glucan metabolism,* respectively. Dynamic models were fitted to the expression profiles of these representative genes. Figure 9 visualizes the

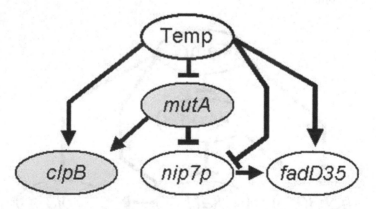

Fig. 9. Structure of the dynamic model C identified by the model fit shown in Figure 10. Here, the set of representative genes selected using GO terms are *mutA* (mutanase, alpha-1,3-glucanase), *clpB* (Clp protease, involved in protein folding), *nip7p* (60S ribosome subunit biogenesis protein) and *fadD35* (long-chain-fatty-acid—CoA ligase). Symbols as in Figure 5.

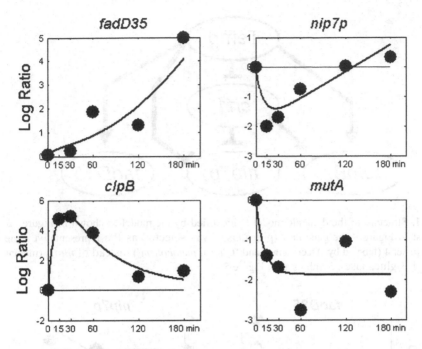

Fig. 10. Measured (•) expression kinetics (log-ratios) for the genes selected as representatives of the four clusters by GO term analysis and kinetics simulated by the model C shown in Figure 9. Model fit error: *mse* = 5.9.

network structure of the model C. The results of model simulation are shown in Figure 10. By resampling with randomized data (standard deviation σ = 0.1) all seven links shown in Figure 9 and the two self regulation were found in more than 99% of the reconstructed models, i.e. they are very stable.

Model D

Gene *mutA* whose expression data are erratic as shown in Figure 10 was substituted by gene *crf1*, which was ranked in Table 4 as the second representative candidate for cluster four. Both genes are coding for glucanases. Figure 11 visualizes the network structure of the model D. By resampling with randomized data (standard deviation σ = 0.1) all six links shown in Figure 11 were found in more than 90% of the reconstructed models. Only the self regulation term of *nip7p* was not stable at this percentage level (i.e., it was found only in 70% of the reconstructed models). The results of model simulation are shown in Figure 12. Models C and D have different structures: In contrast to model D, model C has a link from *nip7p* to *fadD35*, whereas model D has a self-regulation term for *nip7p*, which does not apply to model C. Model D is superior to model C, because the fit error of model D is smaller (*mse* = 5.0) than that of model C (*mse* = 5.9) and both models have the same number of model parameters (*N* = 9).

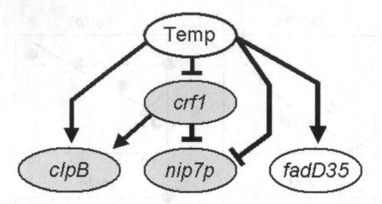

Fig. 11. Structure of the dynamic model D identified by the model fit shown in Figure 12. In contrast to Figure 9 the gene *crf1* (glucanase) was selected as the representative gene of gene cluster 4 (labeled by 'Decreasing' and '*Glucan metabolism*') instead of *matA* (mutanase, alpha-1,3-glucanase). Symbols as in Figure 5.

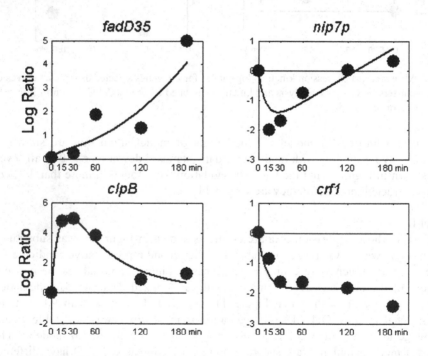

Fig. 12. Measured (•) expression kinetics (log-ratios) for the genes selected as representatives of the four clusters by GO term analysis and kinetics simulated by the model D shown in Figure 11. In opposition to Figure 10 the gene *crf1* was used as representative for cluster four. Model fit error: *mse* = 5.0.

4 Conclusion

After temperature shift from 30 °C to 48 °C four clusters of gene expression profiles were found characterized by increasing and decreasing kinetics as well as by the traversal of temporal minima or maxima. Genes related to the peroxisomal part are mainly up-regulated, whereas the majority of genes assigned to the sterol and glucan metabolism are down-regulated. Most of the genes coding for heat-shock proteins and chaperones (mediating protein folding) show temporal maxima in their expression profiles, whereas profiles of genes coding for ribosomal proteins and proteins involved in rRNA processing and ribosome assembly, pass through temporal minima.

The dynamic modeling was performed for four selected transcripts, which have to be representative for the four clusters. For the selection of cluster-representative genes we applied different criteria. Several model structures (A, B, C and D) were constructed for four sets of representative genes. The different structures have some common features: As an early temperature shift response, the expression of genes involved in the metabolism of sterols and glucanes, i.e., constituents of the cell membrane and cell wall, respectively, as well as the expression of ribosomal proteins and rRNA processing enzymes were repressed. The remodeling of the cell wall in response to heat shock and its regulation by a heat shock transcription factor was shown for the yeast *Saccharomyces cerevisiae* [17]. Genes coding for heat shock proteins and chaperones are primarily induced after the temperature shift, but later their expression is reduced in connection with the repression of the synthesis of cell wall and membrane constituents.

The fit error of model B was the smallest followed by model A. Gene *kat1* used for model B as well as gene *fadD35* used for models C and D are coding for proteins, which are involved in the β-oxidation of fatty acids in the peroxisomes. Gene *kat1* is a peroxisomal protein, but is not involved in the stress response to reactive oxygen species. For this reason we prefer the model A for detailed discussion:

Genes involved in oxidative defense reactions, such as catalase *cat2* and genes encoding for superoxide dismutases, are up-regulated. These genes have been discussed as potential virulence determinants in *A. fumigatus* [18]. In comparison, genes coding for enzymes involved in the biosynthesis of the cell membrane component ergosterol are down-regulated. In general, sterols such as the fungal ergosterol influence membrane fluidity and membrane permeability by decreasing the fluidity of the liquid crystalline phase of membranes [19]. In addition, the down-regulated gene *erg11* (cytochrome P-450 sterol 14-alpha-demethylase) is a specific target of antifungal triazoles [20]. Genes coding for heat shock proteins, such as HSP30, are up-regulated transiently, whereas genes coding for ribosomal proteins, such as the fungal allergen RPL3, are down-regulated transiently, i.e. after two hours the up- or down-regulation is diminished. By reverse engineering of the putative gene regulatory network a hypothesis was generated as illustrated by Figure 5. According to this, the relaxation of the expression of genes coding for heat shock protein HSP30 and ribosomal protein RPL3 is caused by an influence of the down-regulated *erg11*. This hypothesis has to be validated, but some data addressing this relationship are available from the literature. It was demonstrated in the yeast *S. cerevisiae* that ergosterol does not directly effect heat shock. However, biological membranes have been implicated as a primary sensor of environmental stress proteins [21] and a protein kinase in *S. cerevisiae* involved in cell cycle control was shown to be positively regulated by trace amounts of ergosterol [22].

There are alternative hypotheses. One of them is, that the expression minimum of genes coding for ribosomal proteins, which was found at 60 minutes after the temperature shift, is causally related to the expression maximum of genes coding for heat shock proteins that was found 30 minutes after the temperature shift. In general, many heat shock proteins are closely connected to the biosynthesis of proteins by functioning as molecular chaperones, i. e. they assist other proteins in achieving proper folding. The significance of a thermotolerant ribosome assembly for the virulence of *A. fumigatus* was discussed by Bhabhra and Askew 2005 [23]. Furthermore, the importance of these proteins in the virulence of pathogenic fungi and their role in the resistance to antifungal drugs has been discussed [24].

Acknowledgement

This work was supported by the German Federal Ministry for Education and Research BMBF (FKZ 0312704D) and the Priority Program 1160 of the Deutsche Forschungsgemeinschaft. The authors would like to thank Lennart Heinzerling for a review of this paper.

References

1. Brakhage, A.A., Langfelder, K.: The molecular biology of *Aspergillus fumigatus*. Annual Review of Microbiol, 56 (2002) 433-455
2. Nierman, W.C. et al.: Genomic sequence of the pathogenic and allergenic filamentous fungus *Aspergillus fumigatus*. Nature, 438 (2005) 1151-1156
3. Ouyang, M., Welsh, W.J., Georgopoulos, P.: Gaussian mixture clustering and imputation of microarray data. Bioinformatics, 20 (2004) 917-923
4. Troyanskaya, O., Cantor, M., Sherlock, G., Brown, P., Hastie, T., Tibshirani, R., Botstein, D. and Altman, R.B.: Missing values methods for DNA microarrays. Bioinformatics, 17 (2001) 520-525
5. Hathaway, R.J., Hu, Y., Bezdek J.C.: Local Convergence of Tri-Level Alternating Optimization. Neural Parallel Sci. Comput, 9 (2001) 19-28
6. Timm, H.: Fuzzy-Clusteranalyse: Methoden zur Exploration von Daten mit fehlenden Werten sowie klassifizierten Daten, Dissertation, University of Magdeburg, Germany, 2002; http://fuzzy.cs.uni-magdeburg.de/~htimm/data/dissertation.pdf
7. Bezdek, J.C., Pal, S.K.: Fuzzy models for pattern recognition: methods that search for structures in data. IEEE Press, New York (1992)
8. Bezdek, J.C., Pal, N.R.: Some new indexes of cluster validity. IEEE Trans. Syst. Man Cybern., B28 (1998) 301-315
9. Bolshakova, N., Azuaje, F.: Cluster validation techniques for genome expression data. Signal Process., 83 (2003) 825-833
10. Möller, U., Radke, D.: A cluster validity approach based on nearest neighbor resampling. International Conference on Pattern Recognition, Hong Kong, August, 2006.
11. http://www.geneontology.org
12. http://www.cadre.man.ac.uk/Aspergillus_fumigatus

13. Yeung, M.K., Tegner, J., Collins, J.J.: Reverse engineering gene networks using singular value decomposition and robust regression. Proc. Natl. Acad. Sci. USA, 99 (2002) 6163-6168

14. Guthke, R., Möller, U., Hoffmann, M., Thies, F., Töpfer, S.: Dynamic network reconstruction from gene expression data applied to immune response during bacterial infection. Bioinformatics, 21 (2005) 1626-1634

15. Paris, S., Wysong, D., Debeaupuis, J.P., Shibuya, K., Philippe, B., Diamond, R.D., Latge J.P.: Catalases of *Aspergillus fumigatus*. Infect Immun, 71 (2003) 3551-3562

16. Mellado, E., Garcia-Effron, G., Buitrago, M.J., Alcazar-Fuoli, L., Cuenca-Estrella, M., Rodriguez-Tudela, J.L.: Targeted gene disruption of the 14-alpha sterol demethylase (cyp51A) in *Aspergillus fumigatus* and its role in azole drug susceptibility. Antimicrob Agents Chemother. 49 (2005) 2536-2538

17. Imazu, H., Sakurai, H.: *Saccharomyces cerevisiae* heat shock transcription factor regulates cell wall remodeling in response to heat shock. Eukaryot. Cell, 4 (2005) 1050-1056

18. Brakhage, A.A.: Systemic fungal infections caused by *Aspergillus* species: epidemiology, infection process and virulence determinants. Curr Drug Targets 6 (2005) 875-886

19. Beney, L., Gervais, P.: Influence of the fluidity of the membrane on the response of microorganisms to environmental stress. Appl. Microbiol. Biotechnol. 57 (2001), 34-42

20. Ferreira, M. E. et al.: The ergosterol biosynthesis pathway, transporter genes, and azole resistance in *Aspergillus fumigatus*. Med. Mycol., 43 Suppl. 1 (2005) S313-S319

21. Swan, T. M., Watson, K.: Stress tolerance in a yeast sterol auxotroph: role of ergosterol, heat shock proteins and trehalose. FEMS Microbiol. Lett., 169 (1998) 191-197

22. Dahl, C., Biemann, H.-P., Dahl, J.: A protein kinase antigenically related to pp60v-src possibly involved in yeast cell cycle control: Positive in vivo regulation by sterol. Proc. Natl. Acad. Sci. USA, 84 (1987) 4012-4016

23. Bhabra, R., Askew, D. S.: Thermotolerance and virulence of *Aspergillus fumigatus*: role of the fungal nucleolus. Med. Mycol., 43 (2005) S87-S93

24. Burnie, J. P., Carter, T. L., Hodgetts, J. S., Matthews, R. C.: Fungal heat-shock proteins in human disease. FEMS Microbiol. Rev., 30 (2006) 53-88

Complexity Measures for Gene Assembly

Tero Harju[1], Chang Li[2], Ion Petre[2,3], and Grzegorz Rozenberg[4,5]

[1] Department of Mathematics, University of Turku,
Turku Center for Computer Science, FIN-20014 Turku, Finland
harju@utu.fi
[2] Department of Computer Science, Åbo Akademi University,
Turku Center for Computer Science, FIN-20520 Turku, Finland
lchang@abo.fi
[3] Academy of Finland
ipetre@abo.fi
[4] Leiden Institute for Advanced Computer Science, Leiden University,
2333 CA Leiden, The Netherlands
rozenber@liacs.nl
[5] Department of Computer Science, University of Colorado,
Boulder, Co 80309-0347, USA

Abstract. The process of gene assembly in ciliates is a fascinating example of programmed DNA manipulations in living cells. Macronuclear genes are split into coding blocks (called MDSs), shuffled and separated by non-coding sequences to form micronuclear genes. Assembling the coding blocks from micronuclear genes to form functional macronuclear genes is facilitated by an impressive in-vivo implementation of the linked list data structure of computer science. Complexity measures for genes may be defined in many ways, including the number of MDSs, the number of loci, etc. We take a different approach in this paper and propose four complexity measures for genes in ciliates, based on the 'effort' required to assemble the gene. We consider: (a) the types of operations used in the assembly, (b) the number of operations used in the assembly, (c) the length of the molecular folds involved, and (d) the length of the shortest possible parallel assembly for that gene.

"One of the oldest forms of life on Earth has been revealed as a natural born computer programmer."
BBC, September 10, 2001.

1 Introduction

Ciliates are very old eukaryotic unicellular organisms that, through evolution, have developed an unusual way of organizing their genome. Each cell has two types of functionally different nuclei - the *macronucleus* is the somatic nucleus, while the *micronucleus* is the germline nucleus. Depending on the species each type of nuclei may be present in many copies in each cell.

The macronuclear genes are very short molecules, e.g., ranging in the S.nova organisms between 200bp and 3700bp, with an average of 2200 bp in length,

K. Tuyls et al. (Eds.): KDECB 2006, LNBI 4366, pp. 42–60, 2007.

see [22], [19], [3], [4]. As a matter of fact, these are the shortest DNA molecules known in Nature, see [20]! On the other hand, micronuclear genome is organized on very long chromosomes (about 120 chromosomes, each with about 10^7 bp in S.nova, see [19]), with coding sequences occupying as little as 2 - 5% of the genome, see, e.g., [3]. During the process of sexual reproduction, ciliates destroy the old macronuclei and transform a micronucleus into a new macronucleus. Ciliates thus have to identify precisely the genetic material and splice it out from the chromosomes. The complexity of the process is profoundly magnified by the fundamentally different organization of the micronuclear and the macronuclear genomes. This process of converting micronuclear genes to their macronuclear form, called *gene assembly*, is especially involved in a family of ciliates called *Stichotrichs* – we concentrate in this paper on this family.

The macronuclear gene is a contiguous DNA sequence, which is placed on its own chromosome, that (with few exceptions only) is not shared with other genes. The same gene in the micronucleus is broken into pieces called *MDSs (macronuclear destined sequences)* that are separated by noncoding blocks called *IESs (internally eliminated sequences)*. Moreover, the order of MDSs may be permuted (with respect to their order in the macronuclear gene), and some of the MDSs may be inverted. Here is where the challenge of gene assembly lies: ciliates have to identify correctly more than 100 000 MDSs in their genome, see [20], assemble them together in the macronuclear (orthodox) order, and eliminate all IESs. We refer to [12], [19], [23] for more details on ciliates and gene assembly.

A hint on how ciliates achieve gene assembly is given by the structure of MDSs. It turns out that ciliates organize their genomic data as *linked lists* in the style used in computer science, see [19]. A short sequence at the end of each MDS is repeated at the beginning of the MDS that should follow it in the orthodox order, thus (in the terminology of computer science) serving as a pointer in a linked list. It is currently believed that ciliates splice together the consecutive MDSs on the common pointers to assemble the gene. The models for gene assembly in Stichotrichs, such as, e.g., [16], [17] and [8], [21], agree on this generic mechanism.

We consider in this paper the *intramolecular* model of [8], [21]. The model is based on three molecular operations: ld, hi, and dlad. In each of these operations, the molecule folds on itself so that two or more pointers get aligned and through recombination two or more MDSs get combined into a bigger composite MDS. The process continues until all MDSs have been assembled.

First operation: ld. In the operation *(loop, direct repeat)-excision*, or ld for short, a pair of pointers flanking an IES guides the excision of this IES as a circular molecule, as illustrated in Fig. 1. The DNA molecule folds on itself so that the two pointers can get aligned, after which the IES is excised through recombination. As a result, two MDSs get joined and form a bigger composite MDS. It is crucial to note that the excised molecule is an IES (closed into a circular form) and so it does not contain any coding blocks – therefore it is not required to participate anymore in the gene assembly process.

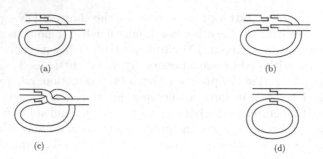

Fig. 1. Illustration of the ld-rule

Second operation: hi. The operation *(hairpin, inverted repeat)-excision/reinsertion,* or hi for short, is applicable to a molecule containing a pair of pointers where one pointer is the inversion of the other. This is illustrated in Fig. 2. The molecule folds on itself forming a hairpin so that the two copies of the pointer can get aligned with the same polarity, thus facilitating the recombination. Through recombination, the sequence between the two occurrences of the pointer is inverted. One may also note that as a result of applying hi, two MDSs are joined together into a bigger composite MDS, while two IESs are joined together into a bigger noncoding block (a bigger composite IES).

Fig. 2. Illustration of the hi-rule

Third operation: dlad. The operation *(double loop, alternating direct repeat)-excision/reinsertion,* or dlad for short applies to a DNA molecule containing two pairs of pointers where the segments delimited by the pairs of pointers overlap with each other. This is illustrated in Fig. 3. The molecule folds into two loops so that the two copies of the first pointer align with each other in one loop, and the two copies of the second pointer align with each other in the other loop. Thus, the molecule is in position for two recombinations. As a result of this double recombination, two sequences are translocated; several MDSs are joined together into bigger composite MDSs(see [6] for details).

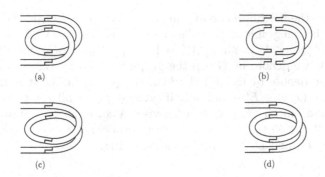

Fig. 3. Illustration of the dlad-rule

2 Definitions

We give in this section some basic notions concerning permutations, strings, and graphs.

For a finite alphabet $\Sigma = \{a_1, \ldots, a_n\}$, we denote by Σ^* the free monoid generated by Σ and call any element of Σ^* a *string*. Let $\overline{\Sigma} = \{\overline{a}_1, \ldots, \overline{a}_n\}$, where $\Sigma \cap \overline{\Sigma} = \emptyset$. For $p, q \in \Sigma \cup \overline{\Sigma}$, we say that p, q have the same *signature* if either $p, q \in \Sigma$, or $p, q \in \overline{\Sigma}$ and we say that they have *different signatures* otherwise. For $p \in \Sigma$, we say that p is an *unsigned letter*, while for $p \in \overline{\Sigma}$, we say that p is a *signed letter*.

Let $\Sigma^{\boldsymbol{\Phi}} = (\Sigma \cup \overline{\Sigma})^*$. For any $u \in \Sigma^{\boldsymbol{\Phi}}$, $u = x_1 \ldots x_k$, with $x_i \in \Sigma \cup \overline{\Sigma}$, for all $1 \leq i \leq k$, we set $\|u\| = \|x_1\| \ldots \|x_k\|$, where $\|a\| = \|\overline{a}\| = a$, for all $a \in \Sigma$. Also, $\overline{u} = \overline{x}_k \ldots \overline{x}_1$, where $\overline{\overline{a}} = a$, for all $a \in \Sigma$.

We say that $u \in \Sigma^{\boldsymbol{\Phi}}$ is a *signed double occurrence string* if for any $p \in \Sigma$, u has either 0, or 2 occurrences from the set $\{p, \overline{p}\}$. In case u has two occurrences from the set $\{p, \overline{p}\}$, we say that p is a *positive letter* in u if the two occurrences have different signatures, and we say that p is a *negative letter* in u if the two occurrences have the same signature. We say that letters p and q, $p \neq q$, *overlap* in u if $u = u_1 p u_2 q u_3 p u_4 q u_5$, for some $u_i \in \Sigma^{\boldsymbol{\Phi}}$, $1 \leq i \leq 5$.

For two signed double occurrence strings, we say that v is a *substring* of u, denoted $v \leq u$, $u = u_1 v u_2$, for some strings u_1, u_2. We say that the signed double occurrence string u is *elementary* if u has no substring v with v a signed double occurrence string.

A *permutation* π over alphabet Σ is a bijection $\pi : \Sigma \to \Sigma$. Fixing the order relation $(a_1, a_2, \ldots, a_m)$ over Σ, we often denote π as the string $\pi(a_1) \ldots \pi(a_m) \in \Sigma^*$. A *signed permutation* over Σ is a string $\psi \in \Sigma^{\boldsymbol{\Phi}}$, where $\|\psi\|$ is a permutation over Σ.

A *signed graph* is a triple $G = (V, E, \phi)$, where V is a finite set of *vertices*, $E \subseteq V \times V$ is the set of (undirected) *edges*, with the property that $(x, y) \in E$ if and only if $(y, x) \in E$, and $\phi : V \to \{+, -\}$ is the *signature function*. We say that vertex $p \in V$ is positive if $\phi(p) = +$ and it is *negative* otherwise. For all $p \in V$, we denote by $N_G(p)$ the neighborhood of p in G, i.e., $N_G(p) = \{q \in V \mid (p, q) \in E\}$. For $V' \subseteq V$, the subgraph of G *induced* by V' is $G_V = (V', E', \phi')$, where $E' = \{(p, q) \in E \mid p, q \in V'\}$ and $\phi' : V' \rightarrow \{+, -\}$, $\phi'(p) = \phi(p)$, for all $p \in V'$.

For all $p \in V$ we denote by $G - p$ the graph induced by the set of vertices $V \setminus \{p\}$. We also denote by $\mathrm{loc}_p(G)$ the *local complement* of G at p: $\mathrm{loc}_p(G) = (V, E', \phi')$, where $(x, y) \in E'$ if and only if $(x, y) \notin E$, for all $x, y \in N_G(p)$, and $(x, y) \in E'$ if and only if $(x, y) \in E$ otherwise. Also, $\phi'(x) = +$ if and only if $\phi(x) = -$, for all $x \in N_G(p)$, and $\phi'(x) = \phi(x)$, otherwise.

We denote by C_4 and D_4 the graphs shown in Fig. 4.

(a)   (b)

Fig. 4. (a) The square C_4; (b) the diamond D_4

With any signed double occurrence string u over alphabet Σ, we associate a signed graph $G_u = (V_u, E_u, \phi_u)$ as follows: $V_u = \{p \in \Sigma \mid p$ or $\overline{p}$ occurs in $u\}$, $E_u = \{(p, q) \mid p$ and q overlap in $u\}$, and $\phi_u(p) = +$ if and only if p is a positive letter in u. The graph G_u is called the *overlap graph* of u.

For $k \geq 2$ we will use throughout the paper the alphabets $\Sigma_k = \{1, \ldots, k\}$ and $\Delta_k = \{2, \ldots, k\}$.

3   Three Models for Gene Assembly

The intramolecular model for gene assembly, [8], [21], has been formalized on several levels of abstraction. The structures of genes can be represented as: signed permutations, MDS descriptors, signed double occurrence strings, or signed overlap graphs. Consequently, the process of gene assembly can be formalized through processing of strings, or through processing of graphs. As it turns out, all these levels of abstraction are equivalent as far as the modeling of gene assembly is concerned, see [6] for a detailed discussion on model forming. Nevertheless, different levels of abstraction prove more suitable (more elegant or technically simpler) for different research topics.

In this paper we consider issues dealing with formalization of the gene assembly on the level of string permutation, signed double occurrence strings, and signed graphs. We present briefly these three abstraction levels referring to [6] for more details.

For any gene γ having k MDSs, $k \geq 1$, we may associate a signed permutation in the following way: associate to the MDS M_i letter i, $1 \leq i \leq k$, and to its inversion $\overline{M}_i$ the signed letter $\overline{i}$. Thus the signed permutation associated to the MDS sequence $M_3 \overline{M}_1 M_2$ is simply the signed permutation $3\overline{1}2$.

We may also associate a signed double occurrence string with any gene (more generally, to any sequence of MDSs), simply by writing its sequence of pointers. Thus, given a sequence of k MDSs, we associate with each MDS M_i, $2 \leq i \leq k-1$, the string consisting of its incoming and outgoing pointers: $i(i+1)$. With $\overline{M}_i$ we associate string $\overline{(i+1)}\overline{i}$. The first and the last MDS are special because they contain only one pointer each, and moreover we mark the beginning of the first MDS by the beginning marker, and the end of the last MDS by the end marker. In our coding of these MDSs, we ignore the beginning and the end markers – thus, with M_1 we associate string 2 and with $\overline{M}_1$ string $\overline{2}$. Similarly, with M_k we associate string k and with $\overline{M}_k$ string $\overline{k}$, Consequently, with the MDS sequence $M_3\overline{M}_1M_2$ we associate string $3\overline{2}23$. Also, with the MDS sequence $M_2\overline{M}_4M_1M_3$ we associate string $23\overline{4}234$.

On a higher level of abstraction, we may associate a graph with a sequence of MDS in the following way. If u_γ is the string associated with gene γ, then G_γ is the signed overlap graph of u_γ, as defined in Section 2. Thus, the graph associated with the MDS sequence $M_3\overline{M}_1M_2$ consists of positive vertex 2, adjacent to negative vertex 3.

The molecular operation ld, hi, dlad are modeled by string rewriting rules as follows (we will use the notation ld, hi, dlad also for the string rules, but this should not lead to confusion).

Let u be a signed double occurrence string over alphabet Δ_k.

1. For all $p \in \Delta_k \cup \overline{\Delta}_k$, ld_p is defined as follows:

$$\mathsf{ld}_p(uppv) = uv,$$

where $u, v \in \Delta_k^{\circledast}$.

2. For all $p \in \Delta_k \cup \overline{\Delta}_k$, hi_p is defined as follows:

$$\mathsf{hi}_p(upv\overline{p}w) = u\overline{v}w,$$

where $u, v \in \Delta_k^{\circledast}$.

3. For all $p \in \Delta_k \cup \overline{\Delta}_k$, $\mathsf{dlad}_{p,q}$ is defined as follows:

$$\mathsf{dlad}_{p,q}(u_1pu_2qu_3pu_4qu_5) = u_1u_4u_3u_2u_5,$$

where $u_i \in \Delta_k^{\circledast}$, for all $1 \leq i \leq 5$.

We say that a composition ϕ of ld, hi, and dlad operations is a *reduction strategy* for string u if $\phi(u) = \Lambda$.

Example 1. Let $u = 3\overline{5}265473672\overline{4}88$, then ld_8 is applicable to u: $\mathsf{ld}_8(u) = 3\overline{5}2654736724$. Also, hi_4 and $\mathsf{dlad}_{5,6}$ are applicable to u: $\mathsf{hi}_4(u) = 3\overline{5}265\overline{2}7\overline{6}3788$, and $\mathsf{dlad}_{3,2}(u) = 6765465\overline{5}488$.

The corresponding operations for signed graphs are defined as follows (again, also for the graph rules we will use the same notation ld, hi, and dlad). Let $G = (V, E)$ be a signed graph.

1. For all $p \in V$, ld_p is applicable to G if and only if p is an isolated negative vertex in G. When applicable, $\mathsf{ld}_p(G) = G - p$.
2. For all $p \in V$, hi_p 4 is applicable to G if and only if p is an positive vertex in G. When applicable, $\mathsf{hi}_p(G) = \mathsf{loc}_p(G) - p$.
3. For all $p, q \in V$, $\mathsf{dlad}_{p,q}$ is applicable to G if and only if p and q are adjacent negative vertices in G. When applicable, $\mathsf{dlad}_{p,q}(G) = (V \setminus \{p, q\}, E')$, where E' is obtained from E by complementing the edges that join vertices in $N_G(p)$ with vertices in $N_G(q)$. This means that $(x, y) \in (E' \setminus E) \cup (E \setminus E')$ if and only if

$$x \in N_G(p) \setminus N_G(q) \text{ and } y \in N_G(q), \text{ or}$$
$$x \in N_G(q) \cup N_G(q) \text{ and } y \in (N_G(p) \setminus N_G(q)) \cup (N_G(q) \setminus N_G(p)), \text{ or}$$
$$x \in N_G(q) \setminus N_G(p) \text{ and } y \in N_G(p).$$

We say that a composition ψ of ld, hi, and dlad operations is a *reduction strategy* for graph G if $\psi(G) = \emptyset$.

Example 2. The signed overlap graph associated to string u in Example 1 is depicted in Fig. 5

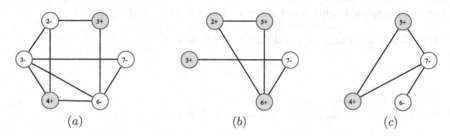

Fig. 5. (a) The graph G_u in Example 2; (b) $\mathsf{hi}_4(G_u)$; (c) $\mathsf{dlad}_{2,3}(G_u)$

Without risk of confusion, for both the string rules and for the graph rules we set $\mathsf{Ld} = \{\mathsf{ld}_p \mid p \geq 2\}$, $\mathsf{Hi} = \{\mathsf{hi}_p \mid p \geq 2\}$, and $\mathsf{Dlad} = \{\mathsf{dlad}_{p,q} \mid p, q \geq 2, p \neq q\}$.

Note that the process of gene assembly, and all its formalizations, as sorting permutations, reducing strings, or reducing graphs, are non-deterministic. We illustrate this in the following example.

Example 3. Consider a double occurrence string $u = 562324573467$. There are at least two reduction strategies for u: $\phi_1 = \mathsf{ld}_7 \circ \mathsf{ld}_4 \circ \mathsf{dlad}_{5,6} \circ \mathsf{dlad}_{2,3}$ and $\phi_2 = \mathsf{ld}_5 \circ \mathsf{ld}_6 \circ \mathsf{dlad}_{2,3} \circ \mathsf{dlad}_{4,7}$. Indeed,

$$\phi_1(u) = (\mathsf{ld}_7 \circ \mathsf{ld}_4 \circ \mathsf{dlad}_{5,6})(56457467) = (\mathsf{ld}_7 \circ \mathsf{ld}_4)(7447) \doteq \mathsf{ld}_7(77) = \Lambda,$$

$$\phi_2(u) = (\mathsf{ld}_5 \circ \mathsf{ld}_6 \circ \mathsf{dlad}_{2,3})(56232635) = (\mathsf{ld}_5 \circ \mathsf{ld}_6)(5665) = \mathsf{ld}_5(55) = \Lambda.$$

Note that the two strategies have the same number of ld operations, albeit applied to different pointers.

The following result, adapted from [6] provides an invariant for all sorting strategies of a given string.

Theorem 1 ([6]). *Let u be a signed double occurrence string and ϕ_1, ϕ_2 two reduction strategies for u. Then ϕ_1 and in ϕ_2 contain the same number of* ld *operations.*

4   First Complexity Measure: The Minimal Subset of Operations Sufficient for Gene Assembly

We introduce in this section our first measure of gene complexity in terms of the smallest set of (types of) operations that are capable to assemble a given gene. Our formalism in this section will be that of signed double occurrence strings. Note that a similar presentation may also be done in terms of signed graphs, see [5].

The concept of (gene) complexity here is the following. For a given string x, consider reduction strategies φ for x, and take the set $S_\varphi \subseteq \{\mathsf{Ld}, \mathsf{Hi}, \mathsf{Dlad}\}$ of those types of operations that are used in φ. We say that S_φ is a *reduction set* for x.

Example 4. Note that a string may have several reduction sets. For instance, if $u = 2\bar{3}2434$, then $\varphi_1(u) = \mathsf{dlad}_{3,4} \circ \mathsf{hi}_2$ is a reduction strategy for u: $\varphi_1 = \mathsf{dlad}_{3,4}(3434) = \Lambda$. Thus, $\{\mathsf{Hi}, \mathsf{Dlad}\}$ is a reduction set for u. However, $\{\mathsf{Hi}\}$ is also a reduction set for u. Indeed, $\varphi_2 = \mathsf{hi}_2 \circ \mathsf{hi}_4 \circ \mathsf{hi}_3$ is a reduction strategy for u: $\varphi_2(u) = (\mathsf{hi}_2 \circ \mathsf{hi}_4)(2\bar{4}24) = \mathsf{hi}_2(2\bar{2}) = \Lambda$.

We say that a set $S \subseteq \{\mathsf{Ld}, \mathsf{Hi}, \mathsf{Dlad}\}$ is a *minimal reduction set* for X, if for any $T \subseteq S$, where T is a reduction set for X, we have $T = S$.

As we will observe at the end of this section, a string X has a unique minimal reduction set. Anticipating this result, the following notion of complexity is well defined.

Definition 1. *The complexity $\mathcal{C}_1(X)$ of a signed double occurrence string X is a minimal reduction set of X.*

To prove the result announced above, we need to consider for every $S \subseteq \{\mathsf{Ld}, \mathsf{Hi}, \mathsf{Dlad}\}$, what are the strings with S as a reduction set. The first complete characterization was given in [5] in the case of realistic strings. The results were then extended to signed double occurrence strings in [1]; the characterizations in [1] are based in part on a notion of break point graphs. For simplicity, we only consider here the case of elementary strings and the approach in [5].

Theorem 2 ([5]). *Let u be an elementary string.*

(i) $\{\mathsf{Ld}\}$ is a reduction set for u if and only if u contains neither overlap, nor signed letters.

(ii) $\{\mathsf{Ld}, \mathsf{Hi}\}$ is a reduction set for u if and only if $|u| \leq 2$ or u contains at least one positive pointer.

(iii) {Ld, Dlad} *is a reduction set for u if and only if u contains no signed letters.*

(iv) {Ld, Hi, Dlad} *is a reduction set for any signed double occurrence string.*

We omit in this paper the characterizations of the strings with {Hi}, {Dlad}, or {Hi, Dlad} as reduction sets. Such characterization have been given in [5].

Example 5.   (a) The R_1 gene of S.nova is described by the MDS sequence $M_1 M_2 M_3 M_4 M_5 M_6$. Its associated string 2233445566 has {Ld} as a minimal reduction set.

(b) {Dlad} is a minimal reduction set for string 2323.

(c) {Hi} is a minimal reduction set for string $23\bar{2}3$.

(d) String $2\bar{3}\bar{2}3$ has two reduction strategies: $ld_3 \circ hi_2$ and $ld_2 \circ hi_3$. Thus, {Ld, Hi} is a minimal reduction set for it.

(e) The α-TP gene of S.nova is described by the MDS sequence $M_1 M_3 M_5 M_9$ $M_{11} M_2 M_4 M_6 M_8 M_{10} M_{12} M_{13} M_{14}$. Its associated string 2345691011122345 678910111213131414 has {Ld, Dlad} as a minimal reduction set.

(f) The actin I gene of S.nova is described by the MDS sequence $M_3 M_4 M_6 M_5$ $M_7 M_9 \overline{M}_2 M_1 M_8$. Its associated string $3445675678\overline{3}\overline{2}289$ has {Ld, Hi, Dlad} as a minimal reduction set.

We can now prove the following result.

Theorem 3. *Let $u \neq \Lambda$ be an elementary string and S_1, S_2 two minimal reduction sets for u. Then $S_1 = S_2$.*

Proof. Assume that there is an elementary string $u \neq \Lambda$ with two different minimal reduction sets S_1, S_2. Clearly, $S_1 \nsubseteq S_2$ and $S_2 \nsubseteq S_1$. We then have the following cases:

(i) $S_1 = \{Ld\}$, $S_2 = \{Hi\}$;   (vi) $S_1 = \{Ld, Hi\}$, $S_2 = \{Hi, Dlad\}$;

(ii) $S_1 = \{Ld\}$, $S_2 = \{Dlad\}$;   (vii) $S_1 = \{Ld, Dlad\}$, $S_2 = \{Hi\}$;

(iii) $S_1 = \{Ld\}$, $S_2 = \{Hi, Dlad\}$;   (viii) $S_1 = \{Ld, Dlad\}$, $S_2 = \{Hi.Dlad\}$;

(iv) $S_1 = \{Ld\,Hi\}$, $S_2 = \{Dlad\}$;   (ix) $S_1 = \{Hi\}$, $S_2 = \{Dlad\}$.

(v) $S_1 = \{Ld, Hi\}$, $S_2 = \{Ld, Dlad\}$;

In all cases except (ii) and (vi) we have that one of the reduction sets contains Hi, while the other does not. Consequently, according to one reduction set, u should have at least one signed letter, while according to the other reduction set, u should have none. Thus, $u = \Lambda$, a contradiction.

In Cases (ii) and (vi) u has two reduction strategies: one containing at least one Ld- operation, another containing none. This is a contradiction by Theorem 1.

Corollary 1. *The complexity measure $\mathcal{C}_1$ is well defined.*

5   Second Complexity Measure: Weights Associated with the Assembly Operations

The concept of our second measure of complexity is straightforward: a gene is more "complex" than another if it requires more "effort" to be assembled. The simplest way to measure the "effort" required to assemble a given gene is through counting the number of operations required in the reduction.

Definition 2. *Let u be a signed double occurrence string and φ a reduction strategy for u. We denote by $\mathcal{C}_2^{(1)}(\varphi)$ the number of* ld, hi, *and* dlad *operations in φ. Then the complexity $\mathcal{C}_2^{(1)}(u)$ is defined as:*

$$\mathcal{C}_2^{(1)}(u) = min\{\mathcal{C}_2^{(1)}(\varphi) \mid \varphi \text{ is a reduction strategy for } u\}.$$

Example 6. Consider $u = 2\bar{3}\bar{2}434$ and reduction strategies φ_1, φ_2 for u as given in Example 4. Then $\mathcal{C}_2^{(1)}(\varphi_1) = 2$, while $\mathcal{C}_2^{(1)}(\varphi_2) = 3$. It is easy to see that $\mathcal{C}_2^{(1)}(u) = 2$.

Clearly, to find the complexity $\mathcal{C}_2^{(1)}(u)$ for a given string u, one needs to find the length of a reduction strategy φ for u using maximum number of dlad operations. Indeed, note that ld and hi operations reduce the length of the string by two, while dlad operations reduce the length of the string by four.

Finding the complexity $\mathcal{C}_2^{(1)}(u)$ is easy if $\mathcal{C}_1(u) \neq \{\text{Hi}, \text{Dlad}\}$. Indeed, based on Theorem 1, it is easy to see that in this case, for any two reduction strategies φ and ψ for u, we have $\mathcal{C}_2^{(1)}(\varphi) = \mathcal{C}_2^{(1)}(\psi)$. It is currently unknown how to compute $\mathcal{C}_2^{(1)}(u)$ if $\mathcal{C}_1(u) = \{\text{Hi}, \text{Dlad}\}$.

Considering the molecular model of the dlad operations, with a double fold and two simultaneous recombination, it may sometimes be undesirable to maximize the number of dlad operations as done when computing $\mathcal{C}_2^{(1)}(u)$. A different idea is to associate *weights* with each of ld, hi and dlad and consequently to any reduction strategy. Associating weights to the operations may be done in at least two ways: either by introducing a (fixed) weight for each type of operation, or through variable weights depending on the type of operation and the string to which the operation applies. We illustrate both ideas in the following.

Definition 3. *For any operation $f \in \text{Ld} \cup \text{Hi} \cup \text{Dlad}$, we define $\mathcal{C}_2^{(2)}(f)$ as follows: $\mathcal{C}_2^{(2)}(f) = c_1$, if $f \in \text{Ld}$; $\mathcal{C}_2^{(2)}(f) = c_2$, if $f \in \text{Hi}$; and $\mathcal{C}_2^{(2)}(f) = c_3$, if $f \in \text{Dlad}$, where $c_1, c_2, c_3 \geq 0$. Then for a composition $\varphi = f_k \circ \cdots \circ f_1$, $f_i \in \text{Ld} \cup \text{Hi} \cup \text{Dlad}$, we let $\mathcal{C}_2^{(2)}(\varphi) = \sum_{i=1}^{k} \mathcal{C}_2^{(2)}(f_i)$.*

For a signed double occurrence string u, the complexity $\mathcal{C}_2^{(2)}(u)$ is defined as

$$\mathcal{C}_2^{(2)}(u) = min\{\mathcal{C}_2^{(2)}(\varphi) \mid \varphi \text{ is a reduction strategy for } u\}.$$

Note that if we define $\mathcal{C}_2^{(2)}(f) = 1$, for any $f \in \text{Ld} \cup \text{Hi} \cup \text{Dlad}$, then $\mathcal{C}_2^{(2)} = \mathcal{C}_2^{(1)}$: we only count the number of operations in each strategy.

Example 7. Let $u = 2\bar{3}\bar{2}434$ and let φ_1, φ_2 be reduction strategies for u as in Examples 4 and 6. Let assign the weights as follows: $\mathcal{C}_2^{(2)}(f) = 0$ for $f \in \text{Ld}$, $\mathcal{C}_2^{(2)}(f) = 1$ for $f \in \text{Hi}$, $\mathcal{C}_2^{(2)}(f) = 3$ for $f \in \text{Dlad}$. Then $\mathcal{C}_2^{(2)}(\varphi_1) = 4$ and $\mathcal{C}_2^{(2)}(\varphi_2) = 3$.

A more refined measure of complexity may be introduced depending on the length of the strings "manipulated" by each operation: the length of the string inverted by hi_p, and the length of the strings translocated by $\mathsf{dlad}_{p,q}$. In the case of ld_p, the excised string is always the same, pp and so, for simplicity, we may set the complexity of ld equal to zero. Formally this is defined as follows.

Definition 4. *Let u be a signed double occurrence string.*

(i) *For any operation ld_p applicable to u, we let $\mathcal{C}_2^{(3)}(\mathsf{ld}_p, u) = 0$.*

(ii) *For any operation hi_p applicable to u, we let $\mathcal{C}_2^{(3)}(\mathsf{hi}_p, u) = |u_2|$, where $u = u_1 p u_2 \bar{p} u_3$, for some strings u_1, u_2, u_3.*

(iii) *For any operation $\mathsf{dlad}_{p,q}$ applicable to u, we let $\mathcal{C}_2^{(3)}(\mathsf{dlad}_{p,q}, u) = |u_2| + |u_4|$, where $u = u_1 p u_2 q u_3 p u_4 q u_5$, for some strings u_1, u_2, u_3, u_4, u_5.*

For a reduction strategy $\varphi = f_k \circ \cdots \circ f_1$ for u, $f_i \in \mathsf{Ld} \cup \mathsf{Hi} \cup \mathsf{Dlad}$, we let $\mathcal{C}_2^{(3)}(\varphi, u) = \sum_{i=1}^{k} \mathcal{C}_2^{(3)}(f_i, (f_{i-1} \circ \cdots f_1)(u))$. Then we define the complexity $\mathcal{C}_2^{(3)}(u)$ by

$$\mathcal{C}_2^{(3)}(u) = min\{\mathcal{C}_2^{(3)}(\varphi, u) \mid \varphi \text{ is a reduction strategy for } u\}.$$

Example 8. Let $u = 3445675678 9\bar{3}\bar{2} 289$ be the string associated with the gene actin I in S.nova. Then $\varphi_1 = \mathsf{ld}_6 \circ \mathsf{dlad}_{7,5} \circ \mathsf{ld}_4 \circ \mathsf{hi}_2 \circ \mathsf{hi}_8 \circ \mathsf{hi}_9 \circ \mathsf{hi}_3$ is a reduction strategy for u:

$$u_1 = \mathsf{hi}_3(u) = \bar{9}\bar{8}\bar{7}\bar{6}\bar{5}\bar{7}\bar{6}\bar{5}\bar{4}\bar{4}\bar{2}289,$$
$$u_2 = \mathsf{hi}_9(u_1) = \bar{8}\bar{2}2445675678,$$
$$u_3 = \mathsf{hi}_8(u_2) = \bar{7}\bar{6}\bar{5}\bar{7}\bar{6}\bar{5}\bar{4}\bar{4}\bar{2}2,$$
$$u_4 = \mathsf{hi}_2(u_3) = \bar{7}\bar{6}\bar{5}\bar{7}\bar{6}\bar{5}\bar{4}\bar{4},$$
$$u_5 = \mathsf{ld}_4(u_4) = \bar{7}\bar{6}\bar{5}\bar{7}\bar{6}\bar{5},$$
$$u_6 = \mathsf{dlad}_{7,5}(u_5) = \bar{6}\bar{6},$$
$$u_7 = \mathsf{ld}_6(u_6) = \Lambda.$$

Hence $\mathcal{C}_2^{(3)}(\varphi_1, u) = \mathcal{C}_2^{(3)}(\mathsf{hi}_3, u) + \mathcal{C}_2^{(3)}(\mathsf{hi}_9, u_1) + \mathcal{C}_2^{(3)}(\mathsf{hi}_8, u_2) + \mathcal{C}_2^{(3)}(\mathsf{hi}_2, u_3) + \mathcal{C}_2^{(3)}(\mathsf{ld}_4, u_4) + \mathcal{C}_2^{(3)}(\mathsf{dlad}_{7,5}, u_5) + \mathcal{C}_2^{(3)}(\mathsf{ld}_6, u_6) = 10 + 12 + 10 + 0 + 0 + 2 + 0 = 34$.

Note that $\varphi_2 = \mathsf{hi}_3 \circ \mathsf{hi}_2 \circ \mathsf{dlad}_{8,9} \circ \mathsf{ld}_7 \circ \mathsf{dlad}_{5,6} \circ \mathsf{ld}_4$ is also a reduction strategy for u:

$$v_1 = \mathsf{ld}_4(u) = 35675678 9\bar{3}\bar{2}289,$$
$$v_2 = \mathsf{dlad}_{5,6}(v_1) = 3778 9\bar{3}\bar{2}289,$$
$$v_3 = \mathsf{ld}_7(v_2) = 389\bar{3}\bar{2}289,$$
$$v_4 = \mathsf{dlad}_{8,9}(v_3) = 3\bar{3}22,$$
$$v_5 = \mathsf{hi}_2(v_4) = 3\bar{3},$$
$$v_6 = \mathsf{hi}_3(v_5) = \Lambda.$$

Thus $\mathcal{C}_2^{(3)}(\varphi_2, u) = \mathcal{C}_2^{(3)}(\mathsf{ld}_4, u) + \mathcal{C}_2^{(3)}(\mathsf{dlad}_{5,6}, v_1) + \mathcal{C}_2^{(3)}(\mathsf{ld}_7, v_2) + \mathcal{C}_2^{(3)}(\mathsf{dlad}_{8,9}, v_3) + \mathcal{C}_2^{(3)}(\mathsf{hi}_2, v_4) + \mathcal{C}_2^{(3)}(\mathsf{hi}_3, v_5) = 0$.

A natural question here is: what are the strings with the minimal $\mathbb{C}_2^{(3)}(u)$ complexity? We discuss this issue in the next section, where we consider the simple operations for gene assembly.

6 Third Complexity Measure: Simple Operations

As discussed above, one way to introduce a complexity measure for gene assembly is by considering the length of the molecular folds involved in every step of the assembly. We consider in this section simple versions of ld, hi, and dlad where the operations can only be applied on the shortest possible folds. It is known that Ld∪Hi∪Dlad is a complete model, in the sense that any gene (alternatively: signed permutation, string, or graph) may be assembled in this model, see [7]. It turns out that the simple operations are not complete: there are certain patterns that cannot be assembled through simple operations. Remarkably though, all known micronuclear gene sequences, see [2], can indeed be assembled through simple operations.

The molecular model for simple ld, hi, and dlad was introduced in [11]. Due to lack of space, we only give here a short intuitive presentation, followed by its formalization as rewriting rules for signed permutations. For formalizations on the level of MDS descriptors and signed double occurrence strings we refer to [11].

As observed in Section 3, ld must always be simple – the excised sequences may never contain coding blocks for the assembly to succeed. In simple hi, one only inverts sequences containing *at most one MDS*. Similarly, in simple dlad, the two sequences that are translocated may contain altogether *at most one MDS*. We refer to [11] for details.

As noted in Section 3, when working with signed permutations, we ignore the ld operation and model gene assembly as a process of sorting a signed permutation rather than as a process of pointer elimination. Simple hi and dlad are modeled through the following operations for signed permutations.

1. For each $p \geq 1$, sh_p is defined as follows:

$$\mathsf{sh}_p(xp\ldots(p+i)(\overline{p+k})\ldots(\overline{p+i+1})y) = xp\ldots(p+i)(p+i+1)\ldots(p+k)y,$$
$$\mathsf{sh}_p(x(\overline{p+i})\ldots\overline{p}(p+i+1)\ldots(p+k)y) = xp\ldots(p+i)(p+i+1)\ldots(p+k)y,$$
$$\mathsf{sh}_p(x(p+i+1)\ldots(p+k)(\overline{p+i})\ldots\overline{p}) = x(\overline{p+k})\ldots(\overline{p+i+1})(\overline{p+i})\ldots\overline{p}y,$$
$$\mathsf{sh}_p(x(\overline{p+k})\ldots(\overline{p+i+1})p\ldots(p+i)y) = x(\overline{p+k})\ldots(\overline{p+i+1})(\overline{p+i})\ldots\overline{p}y,$$

where $k > i \geq 0$ and x, y, z are signed strings over Σ_n. Let $\mathsf{Sh} = \{\mathsf{sh}_i \mid 1 \leq i \leq n\}$.

2. For each p, $2 \leq p \leq n - 1$, sd_p is defined as follows:

$$\mathsf{sd}_p(x\,p\ldots(p+i)\,y\,(p-1)\,(p+i+1)\,z) = xy(p-1)p\ldots(p+i)(p+i+1)z,$$
$$\mathsf{sd}_p(x\,(p-1)(p+i+1)yp\ldots(p+i)z) = x(p-1)p\ldots(p+i)(p+i+1)yz,$$

where $i \geq 0$ and x, y, z are signed strings over Σ_n. We also define $\mathsf{sd}_{\bar{p}}$ as follows:

$$\mathsf{sd}_{\bar{p}}(x\,\overline{(p+i+1)}\,\overline{(p-1)}\,y\,\overline{(p+i)}\dots\bar{p}z) = x\,\overline{(p+i+1)}\,\overline{(p+i)}\dots\bar{p}\,\overline{(p-1)}\,yz,$$
$$\mathsf{sd}_{\bar{p}}(x\,\overline{(p+i)}\dots\bar{p}y\,\overline{(p+i+1)}\,\overline{(p-1)}z) = xy\,\overline{(p+i+1)}\,\overline{(p+i)}\dots\bar{p}\,\overline{(p-1)}z,$$

where $i \geq 0$ and x, y, z are signed strings over Σ_n. Let $\mathsf{Sd} = \{\mathsf{sd}_i, \mathsf{sd}_{\bar{i}} \mid 1 \leq i \leq n\}$.

We say that a signed permutation π over a set of integers $\{i, i+1, \dots, i+l\}$ is *sortable* if there are operations $\phi_1, \dots, \phi_k \in \mathsf{Sh} \cup \mathsf{Sd}$ such that $(\phi_k \circ \dots \circ \phi_1)(\pi)$ is a sorted permutation. We say that π is *blocked* if neither an sh operation, nor an sd operation is applicable to π and π is not sorted.

Let $\phi = \phi_k \circ \dots \circ \phi_1$, $\phi_i \in \mathsf{Sh} \cup \mathsf{Sd}$, for all $1 \leq i \leq k$. We say that ϕ is a *strategy* for π if $\phi(\pi)$ is either sorted or blocked. In the former case we say that ϕ is a *sorting strategy*, while in the latter case we say that ϕ is a *unsuccessful strategy* for π.

Example 9. Let $\pi = 243\bar{1}$ be a signed permutation. Then $(\mathsf{sh}_1 \circ \mathsf{sd}_3)(\pi) = \mathsf{sh}_1(234\,\bar{1}) = \bar{4}\,\bar{3}\,\bar{2}\,\bar{1}$, a sorted permutation.

One may introduce "elementary" versions of sh and sd, where only one letter is rewritten in every step, rather than strings as in sh and sd. We introduce them in the following.

3. For each $p \geq 1$, eh_p is defined as follows:

$$\mathsf{eh}_p(x\,p\,\overline{(p+1)}\,y) = x\,p\,(p+1)\,y, \qquad \mathsf{eh}_p(x\,\overline{(p+1)}\,p\,y) = x\,\overline{(p+1)}\,\bar{p}\,y,$$
$$\mathsf{eh}_p(x\,\bar{p}\,(p+1)\,y) = x\,p\,(p+1)\,y, \qquad \mathsf{eh}_p(x\,(p+1)\,\bar{p}\,y) = x\,\overline{(p+1)}\,\bar{p}\,y,$$

where x, y are signed strings over Σ_n. Let $\mathsf{Eh} = \{\mathsf{sh}_p \mid 1 \leq p \leq n\}$.

4. For each $p \geq 1$, $2 \leq p \leq n-1$, ed_p is defined as follows:

$$\mathsf{ed}_p(x\,p\,y\,(p-1)\,(p+1)\,z) = x\,y\,(p-1)\,p\,(p+1)\,z,$$
$$\mathsf{ed}_p(x\,(p-1)\,(p+1)\,y\,p\,z) = x\,(p-1)\,p\,(p+1)\,y\,z,$$
$$\mathsf{ed}_p(x\,\overline{(p+1)}\,\overline{(p-1)}\,y\,\bar{p}\,z) = x\,\overline{(p+1)}\,\bar{p}\,\overline{(p-1)}\,y\,z,$$
$$\mathsf{ed}_p(x\,\bar{p}\,y\,\overline{(p+1)}\,\overline{(p-1)}\,z) = x\,y\,\overline{(p+1)}\,\bar{p}\,\overline{(p-1)}\,z,$$

where x, y, z are signed strings over Σ_n. Let $\mathsf{Ed} = \{\mathsf{sd}_p \mid 1 \leq p \leq n\}$.

Example 10. (a) Let $\pi = 3\,\bar{4}\,\bar{5}\,6\,\bar{1}\,2$. Then $(\mathsf{eh}_1 \circ \mathsf{eh}_6 \circ \mathsf{eh}_4 \circ \mathsf{eh}_3)(\pi) = 345612$ is a sorted permutation.
(b) Let $\pi' = 3456\,\bar{1}\,\bar{2}$. Then π' is not $\mathsf{Eh} \cup \mathsf{Ed}$-sortable. Indeed, no eh or ed operation is applicable to π'.

Lemma 1. *For any signed permutation π, if $\mathsf{eh}_p(\mathsf{ed}_p, \text{resp.})$ is applicable to π, for some p, then $\mathsf{sh}_p(\mathsf{sd}_p, \text{resp.})$ is also applicable to π and $\mathsf{eh}_p(\pi) = \mathsf{sh}_p(\pi)$ $(\mathsf{ed}_p(\pi) = \mathsf{sd}_p(\pi), \text{resp.})$*

Note that Lemma 1 does not hold in the reverse direction: if $\pi = 1\,4\,2\,3$, then $\mathsf{sd}_2(\pi) = 1\,2\,3\,4$, while ed_2 is not applicable to π.

As illustrated by the next example, it turns out that the $\mathsf{Eh} \cup \mathsf{Ed}$-model is *nondeterministic*.

Example 11. Let $\pi = 1\,3\,5\,2\,4$. Note that π has both sorting and non-sorting strategies in the elementary model. Indeed, $(\mathsf{ed}_2 \circ \mathsf{ed}_4)(\pi) = 1\,2\,3\,4\,5$, a sorted permutation. On the other hand, $\pi' = \mathsf{ed}_3(\pi) = 1\,5\,2\,3\,4$ is not sorted and no eh or ed operation is applicable to π'.

Due to nondeterminism, deciding whether a given permutation is Eh-, Ed-, or $\mathsf{Eh} \cup \mathsf{Ed}$-sortable is difficult. A complete answer may be found in [10], based on an involved notion of dependency graph.

The simple model however is different. A permutation may indeed have several different strategies, but they are either all sorting, or all non-sorting. Moreover, [13] also defines a notion of *structure of a permutation* and notes that the results obtained after applying these strategies, though different, *have the same structure*. In this way, deciding whether or not a given permutation π is Sh-, Sd-, or $\mathsf{Sh} \cup \mathsf{Sd}$-sortable is easy: simply apply operations from the desired set in an arbitrary order; if the final blocked permutation is sorted, then the answer is 'yes', otherwise the answer is 'no': there are no sorting strategies for π.

Example 12.   (a) The permutation $\pi_1 = 4\overline{6}71\overline{2}35$ has several sorting strategies. Here are some of them:

$$\pi_1^{(1)} = \mathsf{sd}_5 \circ \mathsf{sh}_6 \circ \mathsf{sh}_1(\pi_1) = 4567123,$$

$$\pi_1^{(2)} = \mathsf{sd}_5 \circ \mathsf{sh}_6 \circ \mathsf{sh}_2(\pi_1) = 4567123,$$

$$\pi_1^{(3)} = \mathsf{sd}_4 \circ \mathsf{sh}_6 \circ \mathsf{sh}_2(\pi_1) = 6712345,$$

$$\pi_1^{(4)} = \mathsf{sh}_6 \circ \mathsf{sh}_1 \circ \mathsf{sd}_4(\pi_1) = 6712345.$$

(b) The permutation $\pi_2 = 13685724$ has several unsuccessful strategies. Here are some of them:

$$\pi_2^{(1)} = \mathsf{sd}_2 \circ \mathsf{sd}_7(\pi_2) = 12367854,$$

$$\pi_2^{(2)} = \mathsf{sd}_2 \circ \mathsf{sd}_6(\pi_2) = 12385674,$$

$$\pi_2^{(3)} = \mathsf{sd}_3 \circ \mathsf{sd}_7(\pi_2) = 18567234,$$

$$\pi_2^{(4)} = \mathsf{sd}_3 \circ \mathsf{sd}_6(\pi_2) = 18567234.$$

7   Fourth Complexity Measure: Parallelism

The previous three measures of complexity all deal with *sequential* compositions of operations leading to the assembly of a given gene. We introduce in this section a fourth measure of complexity dealing with more general *parallel assemblies* of genes.

A systematic study of parallelism for gene assembly has been initiated in [15]. We only consider in this paper a graph-based presentation of parallelism, although a string-based study is also possible, see [15].

Intuitively, a set of operations can be applied in parallel to a gene pattern if only if each operation's applicability is independent of the other's. In other words, a number of operations can be applied in parallel to a gene pattern if they can be (sequentially) applied in any order to that gene pattern. Note that this is consistent with how parallelism and concurrency are defined in computer science.

E.g., the C_2 gene of S.nova described by the MDS sequence $M_1 M_2 M_3 M_4$ requires three Ld operations. The three Lds can be applied independently of each other and so, they can be applied in parallel. Also, the micronuclear gene R_1 of S.nova described by the MDS sequence $M_1 M_2 M_3 M_4 M_5 M_6$, requires five Ld operations, and all of them can be applied at once. Consequently, its parallel complexity is one, the same as gene C_2.

Parallelism can be defined in terms of signed graphs as follows.

Definition 5 ([15]). *Let $S \subseteq$ Ld $\cup$ Hi $\cup$ Dlad be a set of k rules and let $G = (V, E, \sigma)$ be a signed graph. We say that the rules in S can be applied in parallel to G if for any ordering $\varphi_1, \varphi_2, \ldots, \varphi_k$ of S, the composition $\varphi_k \circ \cdots \circ \varphi_1$ is applicable to G.*

The following result provides a simple criterium for two rules to be applicable in parallel.

Lemma 2 ([15]). *Let $G = (V, E, \sigma)$ be a signed graph and let $\varphi, \psi \in$ Ld $\cup$ Hi $\cup$ Dlad be two rules applicable to G with $\mathsf{dom}(\varphi) \cap \mathsf{dom}(\psi) = \emptyset$.*

(i) If $\varphi \in$ Ld, then φ and ψ can be applied in parallel to G.

(ii) If $\varphi = \mathsf{hi}_p$ with $p \in V$, then φ and ψ can be applied in parallel to G if and only if $N_G(p) \cap \mathsf{dom}(\psi) = \emptyset$.

(iii) If $\varphi, \psi \in$ Dlad, then φ and ψ can applied in parallel to G if and only if the subgraph of G induced by $\mathsf{dom}(\varphi) \cup \mathsf{dom}(\psi)$ is not isomorphic to either C_4 or D_4.

According to the definition, if a set of rules is applicable in parallel to a signed graph, then any composition of these rules is applicable to that graph. This definition does not require that the result of applying different compositions of rules must be the same. However, it can be proved that this is indeed the case.

Lemma 3 ([15]). *If $\varphi, \psi \in$ Ld$\cup$Hi$\cup$Dlad are applicable in parallel to the signed graph G, then $\varphi(\psi(G)) = \psi(\varphi(G))$.*

The general case follows now easily from Lemma 3.

Theorem 4 ([15]). *Let G be a signed graph and let $S \subseteq$ Ld $\cup$ Hi $\cup$ Dlad be a set of rules applicable in parallel to G. Then for any two compositions φ, φ' of the rules in S, $\varphi(G) = \varphi'(G)$.*

Based on Theorem 4 we can now define the notion of parallel complexity.

Definition 6. *Let G be a signed graph. If $S \subseteq \mathsf{Ld} \cup \mathsf{Hi} \cup \mathsf{Dlad}$ is a set of rules applicable in parallel to G, then we say that S is applicable to G and we denote by $S(G)$ the graph obtained as a result of applying to G any sequential composition of the rules in S.*

If $S_1, S_2, \ldots, S_k \subseteq \mathsf{Ld} \cup \mathsf{Hi} \cup \mathsf{Dlad}$ are disjoint sets of rules, $S_i \cap S_j = \emptyset$, for $i \neq j$, we say that $S_k \circ \ldots \circ S_1$ is applicable to G if S_i is applicable to $(S_{i-1} \circ \ldots \circ S_1)(G)$, for all $1 \leq i \leq k$. If $(S_k \circ \ldots \circ S_1)(G) = \emptyset$, then we say that $S_k \circ \ldots \circ S_1$ is a parallel reduction strategy for G. We say that the parallel complexity of $S = S_k \circ \ldots \circ S_1$ is $\mathcal{C}_4(S) = k$.

We define the parallel complexity $\mathcal{C}_4(G)$ of G as follows:

$$\mathcal{C}_4(G) = \min\{\mathcal{C}_4(S) \mid S \text{ is a parallel reduction strategy for } G\}.$$

Example 13. (a) Any discrete graph can be reduced in one parallel step.
(b) The smallest graph with parallel complexity two is shown in Fig. 6(a).
(c) The smallest graph with parallel complexity three is shown in Fig. 6(b).

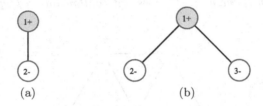

(a)   (b)

Fig. 6. (a) A graph with parallel complexity two; (b) A graph with parallel complexity three

Example 14. Let G be the signed overlap graph associated with actin I gene in S. nova, illustrated in Fig. 7. There are only 6 different maximal parallel strategies to reduce G:

$$S_1 = \{\mathsf{ld}_7, \mathsf{hi}_3\} \circ \{\mathsf{hi}_2, \mathsf{ld}_4, \mathsf{dlad}_{5,6}, \mathsf{dlad}_{8,9}\};$$
$$S_2 = \{\mathsf{ld}_6, \mathsf{hi}_8, \mathsf{hi}_9\} \circ \{\mathsf{hi}_2, \mathsf{hi}_3, \mathsf{ld}_4, \mathsf{dlad}_{5,7}\};$$
$$S_3 = \{\mathsf{ld}_6, \mathsf{hi}_3\} \circ \{\mathsf{hi}_2, \mathsf{ld}_4, \mathsf{dlad}_{5,7}, \mathsf{dlad}_{8,9}\};$$
$$S_4 = \{\mathsf{ld}_7, \mathsf{hi}_8, \mathsf{hi}_9\} \circ \{\mathsf{hi}_2, \mathsf{hi}_3, \mathsf{ld}_4, \mathsf{dlad}_{5,6}\};$$
$$S_5 = \{\mathsf{ld}_5, \mathsf{hi}_3\} \circ \{\mathsf{hi}_2, \mathsf{ld}_4, \mathsf{dlad}_{6,7}, \mathsf{dlad}_{8,9}\};$$
$$S_6 = \{\mathsf{ld}_5, \mathsf{hi}_8, \mathsf{hi}_9\} \circ \{\mathsf{hi}_2, \mathsf{hi}_3, \mathsf{ld}_4, \mathsf{dlad}_{6,7}\}.$$

Note that there are 3060 sequential strategies to reduce this graph (and assemble the gene), see [6] – the reason for this difference is that many sequential strategies coincide modulo commutation of some rules. Those rules may be applied in parallel.

The following problem seems to be difficult: check whether or not a given set of rules can be applied in parallel to a given signed graph. In the next theorem we give a simple criterium in the case when at most two dlad rules are to be applied. Giving a general answer, for an arbitrary number of dlad rules, remains an open problem.

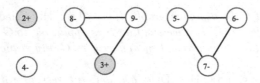

Fig. 7. The signed overlap graph associated with string $3\,4\,4\,5\,6\,7\,5\,6\,7\,8\,9\,\overline{3}\,\overline{2}\,2\,8\,9$, both representing the structure of the micronuclear gene *actin I* in S.nova

Theorem 5 ([15]). *Let G be a signed graph and $S \subseteq \mathsf{Ld} \cup \mathsf{Hi} \cup \mathsf{Dlad}$ a set of rules containing at most two* dlad*'s. Let P be the union of domains of rules in S with $P^+ = \{p \in P \mid \sigma(p) = +\}$, and $P^- = P \setminus P^+$. Then the rules in S can be applied in parallel to G if and only if the following conditions are satisfied:*

(i) *The subgraph induced by P^+ is discrete. Moreover, there is no edge between vertices in P^+ and vertices in P^-.*
(ii) *The subgraph induced by P^- does not contain induced squares C_4 or diamonds D_4.*

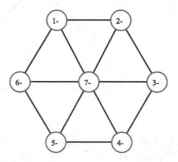

Fig. 8. A negative graph with parallel complexity three

Two conjectures were given in [15] regarding the parallel complexity of graphs. We proposed that any negative graph may be reduced in at most two parallel steps and that any graph may be reduced in at most four parallel steps. Revisiting these conjectures and based on a newly available gene assembly simulator, see [18], we give in the following counterexamples to both these conjectures. It is currently unknown if the parallel complexity of arbitrary graphs is bounded. Several classes of graphs are shown to have bounded parallel complexity in [9].

Example 15. (a) The negative graph G_3 depicted in Fig. 8 has parallel complexity three. (As a matter of fact, an automated search shows that this is a smallest such graph in terms of number of vertices.) Indeed, one three-step parallel strategy for G_3 is $\{\mathsf{Id}_6\} \circ \{\mathsf{dlad}_{5,7}\} \circ \{\mathsf{dlad}_{1,2}, \mathsf{dlad}_{3,4}\}$. Some straightforward analysis shows that no two-step or one-step parallel strategy for G_3 exists.

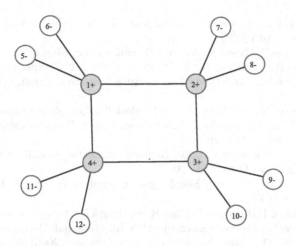

Fig. 9. A graph with parallel complexity five

(b) The graph G_5 depicted in Fig. 9 has parallel complexity 5. One 5-step parallel reduction for G_5 is $\{\mathsf{Id}_8, \mathsf{Id}_{12}\} \circ \{\mathsf{Id}_6, \mathsf{Id}_{10}, \mathsf{hi}_7, \mathsf{hi}_{11}\} \circ \{\mathsf{hi}_5, \mathsf{hi}_9\} \circ \{\mathsf{hi}_2, \mathsf{hi}_4\} \circ \{\mathsf{hi}_1, \mathsf{hi}_3\}$.

Acknowledgments. The authors gratefully acknowledge support by Academy of Finland (TH – project 39802, CL – project 203667, IP – project 108421) and NSF (GR – grant 0121422).

References

1. Brijder, R., Hoogeboom, H.J., Rozenberg, G., Reducibility of gene patterns in cliates using the breakpoint graph, to appear in *Theoret. Comput. Sci* (2006)
2. Cavalcanti, A., Clarke, T.H., Landweber, L., MDS_IES_DB: a database of macronuclear and micronuclear genes in spirotrichous ciliates. *Nucleic Acids Research* **33** (2005) 396–398.
3. Chang, W.J., Bryson, P.D., Liang, H., Shin, M.K., Landweber, L., The evolutionary origin of a complex scrambled gene. *Proceedings of the National Academy of Sciences of the US* **102**(42) (2005) 15149–15154
4. Chang, W.J., Kuo, S., Landweber, L., A new scrambled gene in the ciliate *Uroleptus. Gene* (2006), to appear
5. Ehrenfeucht, A., Harju, T., Petre, I., and Rozenberg, G. (2002) Characterizing the micronuclear gene patterns in ciliates. *Theory of Comput. Syst.* **35** pp 501–519
6. Ehrenfeucht, A., Harju, T., Petre, I., Prescott, D. M., and Rozenberg, G. (2004) Computation in Living Cells: Gene Assembly in Ciliates, Springer
7. Ehrenfeucht, A., Petre, I., Prescott, D. M., and Rozenberg, G., Universal and simple operations for gene assembly in ciliates. In: V. Mitrana and C. Martin-Vide (eds.) *Words, Sequences, Languages: Where Computer Science, Biology and Linguistics Meet*, Kluwer Academic, Dortrecht, (2001) pp. 329–342
8. Ehrenfeucht, A., Prescott, D. M., and Rozenberg, G., Computational aspects of gene (un)scrambling in ciliates. In: L. F. Landweber, E. Winfree (eds.) *Evolution as Computation*, Springer, Berlin, Heidelberg, New York (2001) pp. 216–256

9. Harju, T., Li, C., and Petre, I., Results on parallel reductions of signed overlap graphs, manuscript (2006)
10. Harju, T., Petre, I., Rogojin, V., and Rozenberg, G., Simple operations for gene assembly. In: A. Carbone, N. A. Pierce (eds.) *DNA Computing: 11th International Workshop on DNA Computing*, Lecture Notes in Comput. Sci. **3892** (2006), 96 – 111.
11. Harju, T., Petre, I., and Rozenberg, G., Modelling simple operations for gene assembly. In: J.Chen, N.Jonoska, G.Rozenberg (Eds.) *Nanotechnology: Science and Computation* (2006) 361–376
12. Jahn, C. L., and Klobutcher, L. A., Genome remodeling in ciliated protozoa. *Ann. Rev. Microbiol.* **56** (2000), 489–520.
13. Langille, M., Petre, I. (2006) Simple gene assembly is deterministic. *Fundamenta Informaticae* IOS Press
14. Harju, T., Petre, I., Rogojin, V., and Rozenberg, G. (2006), Simple operations for gene assembly, In: Proceedings of the 11th International Meeting on DNA-based computers DNA11 *Lecture Notes in Computer Science* (2006) Springer
15. Harju, T., Li, C., Petre, I., and Rozenberg, G., Parallelism in gene assemby, In: Proceedings of the 10th International Meeting on DNA-based computers DNA 10, Milan, Italy, *Lecture Notes in Computer Science* **3384** (2005) 140–150
16. Landweber, L. F., and Kari, L., The evolution of cellular computing: Nature's solution to a computational problem. In: *Proceedings of the 4th DIMACS Meeting on DNA-Based Computers*, Philadelphia, PA (1998) pp. 3–15
17. Landweber, L. F., and Kari, L., Universal molecular computation in ciliates. In: L. F. Landweber and E. Winfree (eds.) *Evolution as Computation*, Springer, Berlin Heidelberg New York (2002)
18. Petre, I., Skogman, S. (2006) Gene assembly simulator. http://combio.abo.fi/simulator/simulator.php
19. Prescott, D. M., The DNA of ciliated protozoa. *Microbiol. Rev.* **58**(2) (1994) 233–267
20. Prescott, D. M., DNA manipulations in ciliates. In: W.Brauer, H.Ehrig, J.Karhumäki, A.Salomaa (eds.) *Formal and Natural Computing: essays dedicated to Grzegorz Rozenberg*, LNCS 2300, Springer (2002) 394–417
21. Prescott, D. M., Ehrenfeucht, A., and Rozenberg, G., Molecular operations for DNA processing in hypotrichous ciliates. *Europ. J. Protistology* **37** (2001) 241–260
22. Swanton, M.T., Heumann, J.M., Prescott, D.M., Gene-sized DNA molecules of the macronuclei in three species of hypotrichs: size distribution and absence of nicks. Chromosoma **77** (1980) 217–227
23. Yao, M.C., Fuller, P., Xi, X., Programmed DNA Deletion As an RNA-Guided System of Genome Defense, *Science* 300 (2003) 1581–1584

Learning Relations from Biomedical Corpora Using Dependency Trees

Sophia Katrenko and Pieter Adriaans

Human-Computer Studies Laboratory,
University of Amsterdam,
Kruislaan 419, 1098VA, Amsterdam, The Netherlands
katrenko@science.uva.nl, pietera@science.uva.nl

Abstract. In this paper we address the relation learning problem in the biomedical domain. We propose a representation which takes into account the syntactic information and allows for using different machine learning methods. To carry out the syntactic analysis, three parsers, LinkParser, Minipar and Charniak parser were used. The results we have obtained are comparable to the performance of relation learning systems in the biomedical domain and in some cases out-perform them. In addition, we have studied the impact of ensemble methods on learning relations using the representation we proposed. Given that recall is very important for the relation learning, we explored the ways of improving it. It has been shown that ensemble methods provide higher recall and precision than individual classifiers alone.

1 Introduction

Not only the number of publications in the biomedical domain grows rapidly every year, there are also many approaches proposed to how to handle such amount of data.

These approaches primarily consider such tasks as text mining, information extraction and information retrieval. Information retrieval focuses on the retrieval of the full documents, while the goal of information extraction is to find text fragments relevant to the user need. However, it is often useful to get more fine-grained information, for instance, the list of biomedical instances or relations. Such information might be especially important for the curation of existing resources, such as databases of interactions (Albert et al., 2003).

The paper is organized as follows. We start with the discussion of the related work and problem statement. In Section 3, we present our approach and provide motivation for it. Further, we test our approach on two data sets for the interaction extraction. We report on our results and conclude with the discussion and outlook for the future work.

2 Problem Statement and Related Work

The biggest collection of medical documents is Medline, with 2,000 citations added every week. The large size of this collection makes it impossible to annotate it all by

K. Tuyls et al. (Eds.): KDECB 2006, LNBI 4366, pp. 61–80, 2007.

humans. Consequently, there have been several attempts to create smaller annotated corpora based on Medline, such as Genetag used for the gene/protein named entity recognition (NER) (Tanabe et al., 2005), or MedTag, the corpus comprising Genetag, MedPost and ABGene (Smith et al., 2005). There have also been corpora created with a special purpose to be used by the various challenges, e.g. corpus of the annotated gene-protein relations for the "Genic Interaction Extraction Challenge" (Nédellec, 2005).

In general, the relation learning problem can be seen as a two-step process. First, the relation arguments have to be identified. Further, it is necessary to check whether the relation holds. This setting has also been used for the relation discovery in other domains (Zelenko et al., 2003), moreover, it is often assumed that the arguments have already been found. In this case, the relation learning is reduced to the second step which involves procedures enabling such verification. It has been shown by Bunescu et al. (2005) that provided the correct names of proteins are given, the accuracy of relation discovery is much higher.

The relation learning task can be formulated in the following way:

Definition 1 (Relation learning). *Given a data set D [1] and an n-ary relation Rel with the arguments $X, Y \ldots Z$ find all instances $x \in X, y \in Y, \ldots, z \in Z$ ($x, y, z \in D$), such that $Rel(x, y, \ldots, z)$ holds.*

An example of the relation learning task is given below. In the typical scenario, one starts with the preprocessing (which includes such steps as tokenization and might require some additional analysis depending on the method used). The first step consists of named entity recognition, where all proteins occurring in the sentence are identified. There are three of them, *retinoblastoma*, *RIZ*, and *E1A*. The next step is to detect if there are any relations among them. The correct answer is an interaction between *retinoblastoma* and *RIZ*, while *E1A* does not participate in any interaction.

Input: The retinoblastoma protein binds to RIZ, a zing-finger protein that shares an epitope with the adenovirus E1A protein.
Preprocessing: The| *retinoblastoma* | *protein* | *binds* | *to* | *RIZ* | , | *a* | *zing* − *finger* | *protein* | *that* | *shares* | *an* | *epitope* | *with* | *the* | *adenovirus* | *E1A* | *protein* | . |
Step1:The⟨prot⟩ retinoblastoma ⟨/prot⟩ protein binds to ⟨prot⟩ RIZ ⟨/prot⟩ , a zing-finger protein that shares an epitope with the adenovirus ⟨prot⟩ E1A ⟨/prot⟩ protein .
Step2: The ⟨p1 pair="1"⟩⟨prot⟩retinoblastoma⟨/prot⟩⟨/p1⟩ protein binds to ⟨p1 pair="1"⟩⟨prot⟩RIZ⟨/prot⟩⟨/p1⟩, a zing-finger protein that shares an epitope with the adenovirus ⟨prot⟩E1A⟨/prot⟩ protein.
Output: interaction(retinoblasma, RIZ)

---

[1] Where a data set D can be text, semi-structured data, etc.

An interesting observation has been made by Cohen and Hersh (2005) who considered binary biomedical relations. Although the accuracy of relation extraction for many domains (such as news article extraction) crucially depends on the accuracy of named entity recognition and is equal to the cube[2] of the performance of latter, it seems not to hold in the biomedical domain. The conclusion can be drawn that the surrounding context makes it easier to identify the arguments of a relation in the biomedical domain.

Below, we discuss the relation learning task from the following perspectives: types of relations and approaches.

Types of Relations. Relation learning has received much attention in the past decade but it is nevertheless difficult to compare the results obtained by different research groups. It concerns not only the data sets being used but also the types of relations in question. Often, a certain relation is in focus, such as inhibition (Pustejovsky et al., 2002) or a relation between genes and diseases. The latter is a causal relation which can be formulated as a question "Which gene(s) cause(s) a disease Y?" which in turn can be used for the question answering or information retrieval (Hersh & et al., 2005). This type of relation has been studied by Craven and Kumlien (1999), Ray and Craven (2001). More recently, there has been work on the gene-disease relations carried out by Chun et al. (2006). Contrary to the approach taken by Craven and Kumlien (1999) who have used weakly labeled data, Chun et al. aimed at using a corpus annotated by humans. However, it has been shown that if a gene and a disease co-occur, they are likely to be true positives for the relation extraction. 94% of the correctly identified and co-occurring genes and diseases presented a gene-disease relation. We assume, therefore, that in the discovery of a gene-disease relation, it is necessary to study recall.

However, it is necessary to achieve high recall in other relation learning tasks as well. Some of the relations, such as interactions between genes or genes and proteins are more complex. They can be further divided into groups according to type of interaction, such as interactions expressed by explicit action, binding of the protein on the promoter of a gene, etc. Since the arguments of such relations are genes or proteins, it is important to know whether a given relation is symmetric. The asymmetry of a relation also increases the rate of false positives.

Methods. Most approaches to relation learning fall in one of two categories, either *hand-written* patterns or *learning oriented* approaches. The approaches based on the hand-written (usually pattern-oriented) are usually time-consuming since they often assume use of rules (patterns) written by an expert. Consequently, when such rules are applied to the unseen data, they fail to take into account relations expressed in another way. Although patterns provide a high precision, recall might be much lower (Thomas et al., 2000). In the biomedical

---

[2] Since a relation in question is binary, one needs to identify two arguments and a term identifying a relation itself. It is therefore assumed that the performance of the relation identification equals to the cube of the performance of identifying each of the arguments and a link-word.

domain, it has been proposed to use two types of patterns. The first type is sequential and based on the often occurring sequences of words in a sentence. The second type (Khoo et al., 2000) attempts to account for a syntactic structure of a sentence. Taken the dependency structure of a sentence, which is usually represented as a tree, the patterns in the latter case are subtrees. Such patterns are sometimes referred to as graphical (Khoo et al., 2000). A simpler approach is to consider not the dependency tree as a whole but certain predefined syntactic functions. This idea has been used by Hahn and Romacker (2000),Rinaldi et al. (2004) who have focused on the *subject-verb-object patterns*.

The drawback of the approaches using hand-written patterns is their low recall. Another way to construct such patterns is to use a rule learning algorithm. The performance of the rule learning methods has been studied in detail by Bunescu et al. (2005). The authors have addressed the problems of protein identification and extraction of the protein interactions. For the relation extraction, two approaches have been developed, based on the Rapier rule learning method and on the longest common subsequences. It has been shown that these two approaches outperform hand-written rules.

Contrary to the approaches discussed above, Pustejovsky et al. (2002) and Leroy and Chen (2005) have employed finite state automata to learn relations. When testing their approach on the inhibit-relation, Pustejovsky et al. (2002) have received precision of 90,4% and recall of 58,9%. Particularly interesting approach has been proposed by Bunescu and Mooney (2005) who have studied subsequence kernels for relation extraction. Comparative experiments on the AImed corpus (Bunescu & Mooney, 2005) have revealed that the relation kernel outperforms the approaches based on the longest common subsequences and hand-written rules.

Approaches based on a pure co-occurrence of the biomedical terms are also useful, however, their performance depends on the type of a relation (Stephens et al., 2001). As mentioned above, the co-occurrence of terms denoting diseases and genes is likely to provide evidence for the relation between them. In contrast to a gene-disease relation, a relation between genes is less predictable by the pure co-occurrence of genes in a sentence.

Learning with the Background Knowledge. Since many knowledge resources have been created in the biomedical community in recent years, it is especially valuable to test their impact on the biomedical entity extraction task. Leroy and Chen (2005) have presented a hybrid system integrating linguistic parsing with the existing knowledge sources, such as Gene Ontology, UMLS, and the HUGO nomenclature. They have evaluated 549 relations from Medline abstracts containing p53 gene. In comparison to the relations extracted by parser, the relations provided by a co-occurrence based semantic net Concept Space have been less precise and relevant. However, when adding relations containing terms found in GO and HUGO, precision increases. By approaches such as Leroy's, it has been demonstrated that the knowledge sources can contribute to the protein identification or relation extraction tasks.

3   Approach

The approach we take follows the definition of relation discovery as a two-step process, concentrating on the second step only. We assume that we have already identified the arguments of a relation. In what follows, we present our method and give a motivation from both, linguistic and machine learning perspectives.

Syntactic Information for Relation Learning. In the linguistic tradition, a syntactic structure of a sentence can be presented either by constituency or by dependency analysis (Rastier et al., 2001). The principal distinction between two approaches is the following. By the constituency analysis a sentence is represented by non-overlapping groups of words, whereas dependency is a hierarchical relation where every word in a sentence is linked to a word dominating it. For the sentence in (1), the constituency analysis is given in (2) and the dependency structure is presented on Fig.1. In (2), all words are grouped into noun phrases (NP), verb phrases (VP), prepositional phrase (PP) and pre-verbal adverb phrase (ADVP). It can be noticed that one phrase can contain other phrases, but they never overlap.

(1) Cdc25 can be activated in vitro in a Raf1-dependent manner.
(2) (S (NP Cdc25) (VP can) (VP be (VP activated (ADVP in vitro) (PP in (NP a Raf1-dependent manner))))) (. .))

Unlike constituency analysis, dependency structure does not imply linear ordering and in our view is more appropriate for the relation learning task. At the closer inspection of the sentence (1), it can be found that *Cdc25* and *Raf1* are interacting proteins. The evidence supporting such claim is a word *activated*, so it is possible to rephrase a sentence as 'Raf1 activates Cdc25'. It is, however, difficult to use the constituency structure to detect this relation automatically. In contrast, the root of the dependency tree depicted on Fig.1 already consists of the word *activated*. It is also given that *Cdc25* is a subject of the sentence and that it is in passive voice. Following such analysis, it becomes clear that *Cdc25* is an argument of the binary relation of activation. Assuming activation to be an asymmetric relation, we can conclude that *Cdc25* is a target and *Raf1* is an agent of this relation, or activation(Raf1,Cdc25).

Generally speaking, a root of a dependency tree is often a verb. As shown by Sekimizu et al. (1998), a relation between two or more arguments is also often expressed by verbs. We can therefore conclude that a root of a dependency tree conveys information crucial for relation learning. Furthermore, by examining a parent and children of a given node, one can notice that they constitute a local context important for the argument identification. In the example (1) *dependent*, which is a parent of the word *Raf1* indicates that *Raf1* is an agent and not the target of the interaction.

There are several advantages to considering dependency tree levels. First of all, it is possible to test our hypothesis in order to discover which levels are the most important for relation learning. Selecting tree levels can be considered as a feature engineering step. Moreover, since the final representation is of the attribute-value

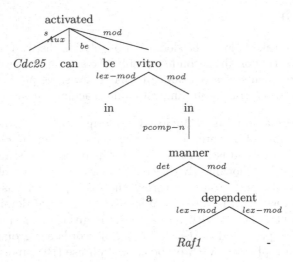

Fig. 1. Dependency structure

type, it is possible to test different machine learning methods. It is of considerable interest to apply ensemble methods (Dietterich, 2000). The experiments for the named entity recognition task have already demonstrated that use of meta-learning improves the accuracy of classification (Sang & Meulder, 2003).

3.1 Defining Levels

We divide all features into two groups, local and global context. To reduce data sparseness, we decided to use lemmas[3] instead of words. A parent of a given node (P) and its two children form a local context. The features of a parent and a child are lemmas and the syntactic function between a node in question and a parent (a child). Since a tree is an acyclic graph, each node has at most one parent but can have more than one child. We limited ourselves to two children, C^1 and C^2.

In addition to a root of a tree (R), a global context consists of a least common subsumer (LCS).

Definition 2 (Least common subsumer (LCS)). *Given two nodes A and B in a dependency tree T, a least common subsumer $LCS(A, B)$ is a node L, such that L is ancestor for both, A and B, and there exist no other node N being an ancestor for A and B, such that L is ancestor of N. There is exactly one LCS for any two nodes in a dependency tree.*

For example, for words a and *Raf1* on Fig.1, the least common subsumer is *manner*. Although such nodes, as *in*, *vitro*, and *activated* are all ancestors of a and *Raf1*, they are not least common subsumers.

---

[3] Lemmas are canonical forms of lexemes, for nouns they usually are nouns in singular, nominative case (such as lemma *dog* for a word *dogs*), for verbs lemmas represent verbs in infinitive (e.g., the lemma *go* for the word *went*).

Algorithm 1

Given text data D and relation $rel(X, Y)$

Parse D to receive a set of dependency trees $S = \cup T_j$, $j = 1, ..., N$

for each relation mention i of $rel(X, Y)$ in T_j **do**

    extract a root R_X of T_j w.r.t. X

    extract a root R_Y of T_j w.r.t. Y

    extract a parent node P_X

    extract a parent node P_Y

    extract the children nodes of X, C_X^1 and C_X^2

    extract the children nodes of Y, C_Y^1 and C_Y^2

    extract a least common subsumer LCS of X and Y

end for

Define a feature set (FS) based on the via the extracted syntactic information

Represent each relation mention i according to FS

Run a chosen learning algorithm on the constructed representation

---

The motivation to include the least common subsumer to the feature set comes from the observation that the arguments of the relation can be located closer to the leaves of a tree and a root in such cases is not sufficiently discriminative. Consider, for instance, a sentence in (3). Here, there are two relation mentions, namely interaction(cwlH,sigma(H)) and interaction(gerE,cwlH). A root of the dependency tree for this sentence is *depended*, which indicates an interaction between *gerE* and *cwlH*. However, it is not sufficient to discover a relation between *gerE* and *sigma(H)*, it is only possible to take into account if a least common subsumer (*dependent*) is used.

(3) Expression of the sigma(K)-dependent cwlH gene depended on gerE.

Table 1. Feature sets

Feature set(FS)	Features
FS1	LCS
FS2	$C_X^1, C_X^2, C_Y^1, C_Y^2, LCS$
FS3	P_X, P_Y, LCS
FS4	C_X^1, C_Y^1, P_X, P_Y
	LCS, R_X
FS5	$C_X^1, C_X^2, C_Y^1, C_Y^2, P_X, P_Y$
	LCS, R_X, R_Y

Some parsers treat a subordinate clause as separate producing not a single tree for a sentence but two. We decided therefore to define two features, one for a root of the first argument (R^1) and second for a root of the second argument(R^2). Table 1 illustrates how the features have been grouped into feature sets given two nodes X and Y and the relation $Rel(X, Y)$.[4]

---

[4] In Table 1, the lower indices correspond to two arguments, X and Y.

Table 2. Feature set for the example on Fig.1

Feature	Cdc25	Raf1
C^1	-	-
C^2	-	-
P	activate_s	dependent_lexmod
LCS	activate	activate
R	activate	activate

When considering the fifth dataset (FS5), the example on Fig.1 can be represented as in Table 2.

Note that we incorporated the syntactic labels into the parent-features P. $Cdc25$ is linked to the word *activated* by the syntactic function s (standing for a subject), while $Raf1$ is connected to *dependent* by *lex-mod* function (standing for a modifier).

4   Experiments

4.1   Datasets

For our experiments, we have used two data sets. One of them is Almed (Bunescu et al., 2005) and the other is a data set created within the "Genic Interaction Extraction" challenge(Nédellec, 2005) (from now, we refer to it as LLL (Learning Language in Logic) data set). The Almed data set consists of the examples of protein-protein interactions. It has been compiled from the 225 Medline abstracts and annotated by the experts. The second data set, LLL, has been created by extracting Medline abstracts on Bacillus subtilis. It also includes annotations created by experts, with the distinction that the focus is on the interactions between genes and proteins. LLL data set consists of 77 sentences and 165 annotated interactions.

4.2   Data Preprocessing

The LLL data set has already consisted of the tokenized sentences accompanied by the syntactic analysis. For parsing, the LLL organizers have used LinkParser whose output has been verified by experts. Besides this, a dictionary of genes and proteins has been provided so we annotated all occurrences of the dictionary items in text as biological entities (in the dictionary, no distinction between genes and proteins has been made).

The second data set has been preprocessed by us. We have used a tokenizer based on the white spaces. We have also found that the present annotation sometimes contains protein tags surrounding the interaction tags as shown in (4). Here, *Ras* is annotated as a protein being an argument of an interaction with *RIN1*. In addition, *Ras binding protein* is also annotated as a protein. As explained below, we have constructed false interactions for the training purpose based on the entities annotated as proteins and not being part of a relation.

We scanned Almed data set and found 14 cases where annotation of a protein included annotation of interactions as its part. While carrying out preprocessing, the external protein tags have been removed.

(4) Human ⟨p1 pair="1"⟩⟨prot⟩RIN1⟨/prot⟩⟨/p1⟩ was first characterized as a ⟨**prot**⟩⟨p2 pair="1"⟩⟨prot⟩Ras⟨/prot⟩⟨/p2⟩ binding protein ⟨/**prot**⟩ based on the properties of its carboxyl-terminal domain.

The data sets we have used provide annotations of the binary interaction relation. The arguments of a relation in LLL always occur in the same sentence, while in Almed they might be in adjacent sentences. We discarded all examples from the latter corpus, which were spanning over several sentences.

In order to obtain the negative interactions, we have followed the closed world assumption. However, it has been used in a different way for each of the data sets. The interactions between proteins in Almed are considered to be symmetric. Therefore, the false positives are created as all pairs of proteins being not arguments of the interaction relation as well as pairs where one of the proteins is an argument of a relation in a given sentence. LLL data set contains interactions between genes and proteins which are treated as instances of an asymmetric relation. Because of this, the false interactions are produced as pairs of biomolecular entities (i.e., proteins or genes) which do not participate in a relation but also those where the arguments of a relation are flipped (e.g., a pair (X,Y) where X and Y are biomolecular entities will be considered a false positive for the true interaction (Y,X)).

After constructing a training set, we received 909 training instances for the LLL data set, 165 of which were positive examples. For the Almed corpus, we obtained 5,106 instances with 1,006 of them being positive examples.

4.3 Parsers

Syntactic analysis is an important step for many text mining tasks in the biomedical domain. In most cases, moving to another domain means necessity of a parser adaptation. Such need is motivated by the domain-specific vocabulary and different stylistic peculiarities (Lease & Charniak, 2005). Lease and Charniak (2005) studied parsing in the biomedical domain and showed that by adapting the Charniak parser, an error rate decreases in 14,2%. Since our approach also crucially depends on the quality of the syntactic analysis, we selected several state-of-the-art parsers to experiment with.

The first parser is LinkParser which has been used by the organizers of LLL data set. It produces relations between pair of words in a sentence, linked by the syntactic function, such as subject or complement. This parser has been adapted to the biomedical domain at MIG Lab. Moreover, the output of the parser was verified by hand, so it can be referred to as the gold standard analysis.

Minipar is another parser, which is freely available.[5] On the Susanne corpus, it achieves 88% precision and 80% recall. Unlike LinkParser, Minipar has not

---

[5] Minipar is available from http://www.cs.ualberta.ca/~lindek/minipar.htm

been trained on the biomedical corpora and it needs to be investigated whether it can be succesfully applied for a given task.

The third syntactic analyzer that we considered is the Charniak parser, whose output can be transformed to the dependencies between a pair of words. This statistical parser was trained on Penn BioIE treebank and part of the Genia treebank and reaches 85% PARSEVAL F-measure when performing 10-fold cross-validation. However, in this case the dependencies do not contain syntactic functions, as in the output of LinkParser. We present the example of the syntactic analysis by all three parsers in Appendix I (Fig.7, Fig.6, Fig.5).

4.4   Data Sets Analysis

To give further motivation for the levels selection, we examined roots of the syntactic trees from LLL data set.

This data set was also processed by 2 researchers in the bioinformatics field whose main intention was to detect all verbs used to express interactions between two biomolecular instances. Such information enables further analysis of the tree levels, namely, it is possible to compare the list of interaction verbs against the words found in a root of the dependency structures. Fig. 8 presents a distribution of verbs selected by our experts which were found in a root of a sentence. 50 out of 77 sentences in the LLL training set contain the selected verbs as roots of the dependency structures. 74,07% of all selected verbs can be found in a root of a dependency tree. Moreover, the verbs which do not occur in a root of a tree, can still be found on other levels. For instance, although *stimulate* and *rely* are not present in a root, they occur in the parents' nodes of the arguments of a relation, as in example (5).

(5) During endospore formation in Bacillus subtilis, the DNA binding protein *GerE* stimulates transcription from several promoters that are used by RNA polymerase containing *sigmaK*.

5   Results and Discussion

The results we present below have been received by 10-fold cross-validation for AImed data set and 5-fold cross-validation for the LLL data set, respectively. We have also used the implementation of the machine learning methods from the Weka toolkit(Witten & Frank, 2005).

Tree Levels. We studied the impact of the feature sets mentioned above on recall and precision. First, we started with a feature set containing the least common subsumer only (FS1). In our view, this feature set is similar to the pattern approaches whose main objective is to find a link (a so-called relation word) between two arguments.

As Table 3 suggests, the least common subsumer often includes the words important for relation learning in the biomedical domain. In the list below, such

Table 3. Least common subsumer: Almed data set

Words (LCS)	Frequency
to be	449
to bind	211
to interact	182
to inhibit	65
to associate with	47
to contain	47
to reveal	43
to induce	42
to include	39
to identify	38
to show	37
to require	36
to regulate	34
to suppress	30
to detect	30
to express	29
to activate	24
to initiate	28
to encode	20
to block	18
to recognize	17
to stimulate	17
to increase	7
to act (as)	3

words are bold-faced. Comparing this list against the list of the verbs identifying relations (Sekimizu et al., 1998), we found that they significantly overlap.

However, using this feature set provides recall and precision, which can likely be further improved. This supports our hypothesis about precompiled list of patterns used for the relation extraction - in most cases, they cannot cover unseen data well. Our results are in line with those reported by Ahmed et al. (2005).

The second feature set, FS2, consists of FS1 and the children of two arguments. As the results on Fig.2 suggest, recall already increases by adding the information about the children.

Using the third feature set containing lemmas from the parent-level provides better results than FS2. We believe it is due to the fact that in many cases proteins are leaves in a tree so the information about the children is missing.

The best performance on LLL has been obtained by employing the third feature set or the fourth set containing all features as defined in Section 3.

As we have mentioned above, in most cases the existing approaches to relation extraction provide considerably high precision but low recall. Our results suggest that the approach we have taken leads to the higher recall and lower precision. It can be concluded that although the information from the dependency tree levels

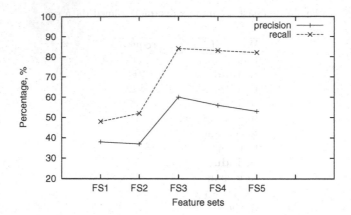

Fig. 2. Precision and recall for different feature sets (Stacking, LLL data set)

helps to find many true positives, the local context is sometimes not sufficient to be able to discriminate between true and false positives.

Ensemble Methods. To test the hypothesis that ensemble methods may improve the overall performance, we have conducted experiments with three ensemble methods, stacking, bagging and AdaBoost. Ensemble of classifiers provide better accuracy if the individual classifiers are diverse and accurate. A classifier is said to be accurate when its error rate is better then random guess (Dietterich, 2000). Bagging and AdaBoost present the ensemble methods manipulating the training examples. The main idea behind such methods lies in generating multiple hypotheses. In the case of bagging, a different subset from the training data is sampled every time a learning algorithm is applied. Bagging and AdaBoost work well for such algorithms as rule learning or decision tree methods which are generally considered to be unstable. Stacking belongs to the method of combining different classification models.

We considered three classifiers of a different nature, BayesNet, Naïve Bayes method and K-nearest neighbor classifier. Bagging and AdaBoost have been applied with BayesNet classifier. The experiment with stacking has been constructed in the following way: BayesNet has been chosen as meta-classifier with NaïveBayes and 1-nearest neighbour classifiers as individual classifiers. The quantitative results of the experiments on both corpora are given in Table 4 and Table 5. In both cases stacking provides much higher recall compared to the supervised methods or other ensemble methods.

(Bunescu & Mooney, 2005) have also used 10-fold cross-validation to test their methods on the Almed corpus. They reported on the performance of their approaches by presenting it as a precision-recall curve. To compare our results on the Almed corpus with the performance of the methods described in (Bunescu & Mooney, 2005), we chose the highest recall received by stacking (68,4%). It corresponds to the precision of 39% on the precision-recall curve which has been received by the subsequence kernel method. We can conclude therefore that our

Table 4. Results on LLL data set

Method	Precision	Recall	F-score
Naïve Bayes	72%	55,6%	62,7%
BayesNet	67,5%	65,4%	66,5%
IB1	**74,5%**	**70,4%**	**72,4%**
Stacking	60,7%	**84%**	70,5%
Bagging	67,5%	67,9%	67,7%
AdaBoostM1	**69%**	71,6%	70,3%

Table 5. Results on AImed data set

Method	Precision	Recall	F-score
Naïve Bayes	48,5%	45,2%	46,8%
BayesNet	46,5%	52,8%	49,5%
IB1	49,7%	49,4%	49,6%
Stacking	45%	**68,4%**	**54,3%**
Bagging	46,6%	52,1%	49,2%
AdaBoostM1	**50,4%**	53,5%	51,9%

method outperforms the subsequence kernel method on **Almed** corpus by 5%. The comparison of the results of the best individual classifier (BayesNet) to the subsequence kernel method demonstrates that it performs equally well. The difference in the performance can be explained by the features used in (Bunescu & Mooney, 2005). In particular, Bunescu and Mooney (2005) have considered sequential information which consisted of the words found between two entities, in front of them and after them. After having defined such features, the authors have restricted themselves to the subsequences of the types mentioned above, where a maximal word length equals 4. According to Bunescu and Mooney (2005), such feature selection leads to less overfitting. In our case, we considered not the common subsequences but the levels from the dependency tree instead. We believe that the selection of levels we have made provides information sufficient for relation learning and constitutes an alternative approach to the method proposed by Bunescu and Mooney (2005).

Comparison of the results received on the **Almed** data set with the performance on the LLL data set (Table 4) demonstrates that we have received much better results on the latter. There are several distinctions between the two data sets. Although LLL data set is much smaller, it contains the syntactic information checked by hand. Some classification errors on the LLL data set can be explained by the asymmetry of the relation between genes and proteins.

Complexity. The feature sets we constructed in our approach are relatively small but able to cover information needed for relation learning. Our approach provides results comparable to that of the state-of-the-art methods. The representation we use results in the training set of the size M*N, where M is the

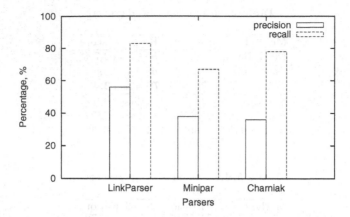

Fig. 3. Precision and recall for different parsers (Stacking, LLL data set)

number of features and N is the number of all potential candidates (in our case, it is a Cartesian product of all proteins found within the same sentence multiplied by the number of sentences). The largest feature set we employ consists of 9 features, thus the maximal size of a training set equals to 9*N. Therefore, the training time depends mostly on a machine learning method used. It is, for example, known that k-nn algorithm is slower than Naïve Bayes, although both of them can be applied to the representation we propose.

Note on Syntactic Analysis. As we already mentioned, the approaches making use of the syntactic structure depend on the accuracy of the parser. Precision and recall of the parsers we used varies and it is likely that some errors in classification are due to the incorrect parsing. To explore if it affects the accuracy of found interactions, we conducted the following experiments. In addition to the LinkParser (whose output was checked by hand), we used Minipar and Charniak parser to analyze the LLL data set. We fixed the feature set (FS4) and used the same machine learning method (Stacking) on the representation received by all three parsers.

As expected, use of the LinkParser output provides the highest precision and recall for the relation learning task (Fig.3). Stacking also boosts recall when used to the Minipar or Charniak parsers' output. However, in both cases precision drops.

We already mentioned in Section 4 that all three parsers give different output. It can be demonstrated on the example in Appendix I. In particular, Minipar produces the syntactic functions for each dependency, whereas the modified output of Charniak parser lacks them. To test whether syntactic functions contribute to the learning task, we conducted another experiment using the same feature set (FS4) and BayesNet classifier with the syntactic functions and without them. To make such comparison fair, we considered LinkParser only. The output of Minipar was not checked by hand so it is not possible to analyze whether the changes in performance are due to the incorrect parsing or to the presence (absence) of the syntactic functions. The results are presented in Fig.4. It can be concluded

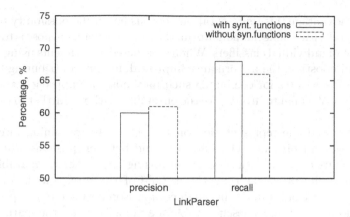

Fig. 4. Syntactic functions

that although removing syntactic functions decrease recall, it does not decrease the overall performance much.

In addition, all three parsers use different sets of the syntactic functions. Sometimes, syntactic functions are grouped or discriminated according to a certain criterion. For instance, such phrases as *transcribed by polymerase* and *activation in prespore* are treated differently by Minipar and LinkParser. While the former introduces a special link *pcomp-n* between a preposition *in* and a noun *prespore*, the latter incorporates preposition in the syntactic function and outputs the relation *comp-in* between *activation* and *prespore*. Such treatment of complements means that the actual training sets differ. LinkParser seems to provide more useful information, because a parent node of *prespore* is *activation* and not *in* as in case of Minipar or Charniak parser.

The performance of current state-of-the- art parsers on the biomedical data has been studied by Grover et al. (2005). The evaluation of a parser has also been done by Rinaldi et al. (2004) who used LT Chunk to obtain verbal and nominal chunks. Nevertheless, the results Rinaldi et al. (2004) have achieved with the correct (verified) syntactic analysis do not differ much from the results received by the parser.

6   Conclusions

In this paper, we have proposed a novel representation for learning relations based on the dependency trees. Learning relations is a difficult task since it is not possible to directly used many well-known machine learning methods using attribute-value representation. This difficulty leads to employing either pattern-based methods or methods capable to use complex representations, such as support vector machines(Zelenko et al., 2003). The representation we use is derived from the complex structures (dependency trees) but it is still attribute-value like representation. Consequently, it can be used for any machine learning method having such representation as its input.

Another advantage this representation gives us lies in the possibility to use ensemble methods. Ensemble methods are methods whose main purpose is to combine the decisions of individual classifiers. When conducted experiments using bagging, stacking and boosting, the performance improved, mainly contributing to higher recall. It also motivates for combining such methods with approaches providing high precision. We plan to investigate the ensemble methods further in our future research.

We have tested the representation on the data sets containing interactions between genes and proteins (LLL data set) and between proteins (Almed data set). The results on both data sets are promising and either comparable to the state-of-art results or better than those. One of the directions in our future research is to carry out a more thorough comparison between our approach and the method proposed by Bunescu and Mooney (2005). Since the feature sets in both approaches are different, we plan to explore whether these two methods can complement each other.

Our experiments suggest that the quality of syntactic analysis is of vital importance when using techniques based on it. As it has been shown, the syntactic analysis verified by the experts is more accurate and this, in turn, leads to higher accuracy of the extracted interactions.

Acknowledgments

The authors would like to thank M. Scott Marshall and Marco Roos for their help in analyzing data sets. Special thanks to Razvan Bunescu who provided us with the Almed data set. Matt Lease made the Charniak's parser trained on the biomedical corpora available. This work was carried out in the context of the Virtual Laboratory for e-Science project (www.vl-e.nl). This project is supported by a BSIK grant from the Dutch Ministry of Education, Culture and Science(OC&W) and is part of the ICT innovation program of the Ministry of the Ministry of Economic Affairs (EZ).

Bibliography

Ahmed, S. T., Chidambaram, D., Davulcu, H., & Baral, C. (2005). Intex: A syntactic role driven protein-protein interaction extractor for bio-medical text. *In Proceedings of the ACL-ISMB Workshop on Linking Biological Literature, Ontologies and Databeses: Mining Biological Semantics* (pp. 54–61). Detroit: Association for Computational Linguistics.

Albert, S., Gaudan, S., Knigge, H., Raetsch, A., Delgado, A., & Huhse, B. (2003). Computer-assisted generation of a protein-interaction database for nuclear receptors. *Molecular Endocrinology*.

Bunescu, R. C., Ge, R., & Kate, R. J. (2005). Comparative experiments on learning information extractors for proteins and their interactions. *Artificial Intelligence in medicine, 33*, 139–155.

Bunescu, R. C., & Mooney, R. J. (2005). Subsequence kernels for relation extraction. *In Proceedings of the 19th Conference on Neural Information Processing Systems*.

Chun, H. W., Tsuruoka, Y., Kim, J. D., Shiba, R., & Nagata, N. (2006). Extraction of gene-disease relations from medline using domain dictionaries and machine learning. *In Proceedings of the 11th Pacific Symposium on Biocomputing*.

Cohen, A. M., & Hersh, W. R. (2005). A survey of current work in biomedical text mining. *Briefings in Bioinformatics, 6(1)*, 57–71.

Craven, M., & Kumlien, J. (1999). Constructing biological knowledge bases by extracting information from text sources. *In Proceedings of the Seventh International Conference on Intelligent Systems for Molecular Biology* (pp. 77–86). Heidelberg, Germany: AAAI Press.

Dietterich, T. G. (2000). Ensemble methods in machine learning. *Lecture Notes in Computer Science, 1857*, 1–15.

Grover, C., Lascarides, A., & Lapata, M. (2005). A comparison of parsing technologies for the biomedical domain. *Natural Language Engineering, 11(1)*, 27–65.

Hahn, U., & Romacker, M. (2000). An integrated model of semantic and conceptual interpretation from dependency structures. *Proceedings of the 18th conference on Computational linguistics* (pp. 271–277). Morristown, NJ, USA: Association for Computational Linguistics.

Hersh, W., & et al. (2005). Trec 2005 genomics track overview. *TREC 2005 meeting*.

Khoo, C. S. G., Chan, S., & Niu, Y. (2000). Extracting causal knowledge from a medical database using graphical patterns. *ACL'00 Proceedings of the 38th Annual Meeting on Association for Computational Linguistics* (pp. 336–343). Morristown, NJ, USA: Association for Computational Linguistics.

Lease, M., & Charniak, E. (2005). Parsing biomedical literature. *In Proceedings of IJCNLP 2005*. Jeju Island, Republic of Korea.

Leroy, G., & Chen, H. (2005). Genescene: An ontology-enhanced integration of linguistic and co-occurrence based relations in biomedical texts. *JASIST, 56*, 457–468.

Nédellec, C. (2005). Learning language in logic - genic interaction extraction challenge. *In Proceedings of the Learning Language in Logic workshop*.

Pustejovsky, J., Castano, J., Zhang, J., Cochran, B., & Kotecki, M. (2002). Robust relational parsing over biomedical literature: Extracting inhibit relations. *Pacific Symposium on Biocomputing*.

Rastier, F., Cavazza, M., & Abeille, A. (2001). *Semantics for descriptions*. Stanford, USA: Center for the Study of Language and Information.

Ray, S., & Craven, M. (2001). Representing sentence structure in hidden Markov models for information extraction. *In Proceedings of the Seventeenth International Joint Conference on Artificial Intelligence* (pp. 1273–1279). Seattle, WA: Morgan Kaufmann.

Rinaldi, F., Schneider, G., & Kaljurand, K. (2004). Mining relations in the genia corpus. *In "Second European Workshop on Data Mining and Text Mining for Bioinformatics", in conjunction with ECML/PKDD 2004*. Pisa, Italy.

Sang, E. T. K., & Meulder, F. D. (2003). Introduction to the conll-2003 shared task: Language-independent named entity recognition. *In Proceedings of CoNLL'2003*.

Sekimizu, T., Park, H., & Tsujii, J. (1998). Identifying the interaction between genes and gene products based on frequently seen verbs in medline abstracts. *Genome Informatics*.

Smith, L. H., Tanabe, L., Rindlesch, T., & Wilbur, W. (2005). Medtag: A collection of biomedical annotations. *In Proceedings of the Joint ACL Workshop and BioLINK SIG (ISMB) on Linking Biological Literature Ontologies and Databases*.

Stephens, M., Palakal, M., Mukhopadhyay, S., Raje, R., & Mostafa, J. (2001). Detecting gene relations from medline abstracts. *In Proceedings of the Sixth Annual Pacific Symposium on Biocomputing (PSB 001)*.

Tanabe, L., Xie, N., L. H. Thom, W. M., & Wilbur, W. J. (2005). Genetag: a tagged corpus for gene/protein named entity recognition. *BMC Bioinformatics, 6(Suppl I):S3*.

Thomas, J., Milward, D., Ouzounis, C., & Pulman, S. (2000). Automatic extraction of protein interactions from scientific abstracts. *In Proceedings of Pacific Symposium on Biocomputing*.

Witten, I. H., & Frank, E. (2005). *Data Mining: Practical machine learning tools and techniques*. San Francisco: Morgan Kaufmann. 2nd edition edition.

Zelenko, D., Aone, C., & Richardella, A. (2003). Kernel methods for relation extraction. *J. Mach. Learn. Res., 3*, 1083–1106.

Appendix I

(6) ykuD was transcribed by SigK RNA polymerase from T4 of sporulation.

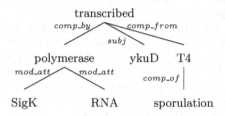

Fig. 5. LinkParser's output

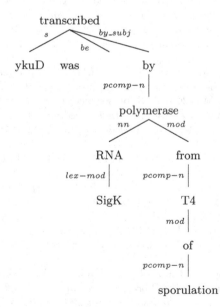

Fig. 6. Minipar's output

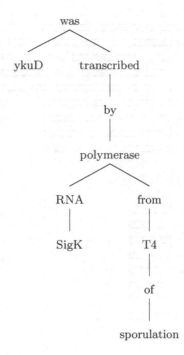

Fig. 7. Charniak parser's output

Table 6. Some of the syntactic functions used in LinkParser

Abbreviation	Name	Example
comp_by	complement(by)	(polymerase, transcribed) 'transcribed by polymerase'
mod_att	modifier(att)	(RNA, polymerase)
comp_of	complement(of)	(T4,sporulation) 'T4 of sporulation'
comp_from	complement(from)	(transcribed, T4) 'transcribed from T4'
subj	subject	(transcribed, ykuD)

Table 7. Some of the syntactic functions used in MiniParser

Abbreviation	Name	Example
s	surface subject	(transcribed, ykuD)
be	be	(transcribed, was)
by-subj	by-subj (for passive voice)	(transcribed, by)
pcomp-n	nominal complement of prepositions	(by, polymerase), (from, T4), (of, sporulation)
nn	complement	(polymerase, RNA)
mod	modifier	(polymerase, from)
lex-mod	modifier	(RNA, SigK)

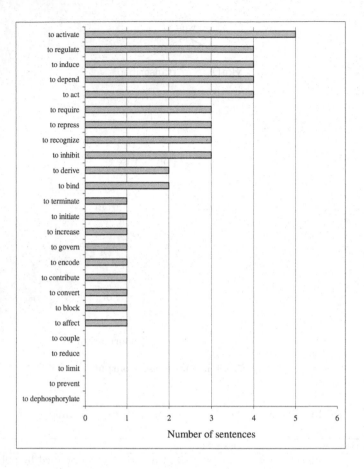

Fig. 8. Distribution of the verbs selected by experts in LLL data set

Advancing the State of the Art in Computational Gene Prediction

William H. Majoros and Uwe Ohler

Center for Bioinformatics and Computational Biology
Institute for Genome Sciences and Policy, Duke University
101 Science Drive, Durham, NC 27708, USA
{bmajoros, uwe.ohler}@duke.edu

Abstract. Current methods for computationally predicting the locations and intron-exon structures of protein-coding genes in eukaryotic DNA are largely based on probabilistic, state-based generative models such as hidden Markov models and their various extensions. Unfortunately, little attention has been paid to the optimality of these models for the gene-parsing problem. Furthermore, as the prevalence of alternative splicing in human genes becomes more apparent, the "one gene, one parse" discipline endorsed by virtually all current gene-finding systems becomes less attractive from a biomedical perspective. Because our ability to accurately identify all the isoforms of each gene in the genome is of direct importance to biomedicine, our ability to improve gene-finding accuracy both for human and non-human DNA clearly has a potential to significantly impact human health. In this paper we review current methods and suggest a number of possible directions for further research that may alleviate some of these problems and ultimately lead to better and more useful gene predictions.

1 Introduction

The growing availability of large quantities of genomic sequence data for both human and non-human species has promoted a renewed interest in purely computational methods for finding protein-coding genes in raw DNA. In the case of vertebrate genomes, the problem has been fairly likened to that of finding the proverbial needle in a haystack, with the additional complication that each needle has an internal structure which also needs to be predicted.

Of the methods which have been investigated for solving this difficult problem, those based on probabilistic models of gene composition and structure have largely come to dominate, with the emphasis in the field now being on the use of hidden Markov models (HMMs) and their various extensions—in particular, those permitting the incorporation of various forms of external evidence such as patterns of evolutionary conservation between related genomes. As the field continues along this track, a number of difficulties have emerged which suggest that the use of purely generative models for heuristic parsing may not be an ideal framework for automated gene prediction.

K. Tuyls et al. (Eds.): KDECB 2006, LNBI 4366, pp. 81–106, 2007.

In particular, the widespread existence of alternative splicing in mammalian genes, the suboptimality of maximum likelihood HMMs for Viterbi parsing, and the lack of efficient discriminative training procedures for stochastic parsers all seem to be conspiring to keep the predictive accuracy of practical gene-finding systems substantially below what is needed by the users of these systems. In the case of biomedical applications, our ability to overcome these limitations may translate into significant impacts on human health.

In this paper we suggest a number of possible directions for further research that may alleviate some of these problems and ultimately lead to better and more useful gene predictions in eukaryotic DNA.

2   Background

2.1   The Problem of Finding and Parsing Eukaryotic Protein-Coding Genes

The human genome comprises 23 chromosomes, each consisting of a single DNA molecule which is in turn formed out of a linear series of *nucleotides*. Nucleotides come in four varieties: *adenine* (A), *cytosine* (C) *guanine* (G), and *thymine* (T). If each nucleotide is denoted by a single letter from the DNA alphabet $\alpha=\{A,C,G,T\}$, the entire genome can then be represented by a sequence of approximately 2.9 billion letters. Embedded within this enormous sequence—at seemingly random intervals— are the actual *genes*, which encode the proteins used by the cell to mediate the building and operation of a complete organism. Expression of a gene begins with its transcription into *messenger RNA* (*mRNA*), which may then be *spliced* by the eukaryotic *spliceosome* to remove stretches of nonfunctional DNA within the gene known as *introns*. The two ends of the mRNA are then specially processed and the message is exported out of the nucleus to await *translation* by a molecular complex called a *ribosome*. This latter process pairs off individual *amino acids* with each triple of nucleotides (called a *codon*) along the message. The concatenation of these amino acids forms a polypeptide which finally folds into a functional protein. In this way, the precise sequence of nucleotides comprising a gene, and the precise way in which that gene's mRNA is spliced, determine the final form of the protein product and thus influence the operation of the cell. Fig. 1 summarizes this process.

The human gene-finding problem is a difficult one for two reasons: (1) the genes comprise less than 2% of our 2.9 billion letter genome, and (2) once a gene is found, the locations of the introns within the gene must be precisely determined before the protein product of the gene may be accurately deduced. The problem is thus one of *parsing*—i.e., partitioning an input sequence into a series of "words" (non-overlapping intervals of various types). The top portion of Fig. 2 shows a sample parse of a DNA sequence; rectangular boxes represent *exons* (non-intronic regions of a gene), the line segments separating pairs of exons represent introns, and the white spaces to the left and right of the gene represent *intergenic* regions.

Shaded portions of exons represent the parts of the gene which are actually translated into amino acids; in typical eukaryotic organisms, only the region between

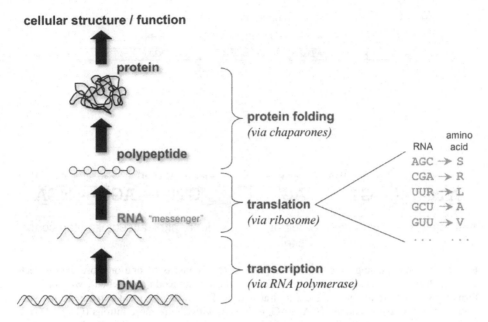

cellular structure / function

Fig. 1. The central dogma of molecular biology: DNA gives rise to RNA messages, which are translated into polypeptides that then fold into functional proteins. Source: Majoros WH, *Methods for Computational Gene Prediction*, Cambridge University Press (forthcoming), reproduced with permission.

the *start codon* (ATG) and the *stop codon* (one of TGA, TAG, or TAA) is translated. Hatched portions of exons in the figure therefore represent *untranslated regions* (*UTRs*), and are generally not predicted by current gene-finding programs (though preliminary work in this direction shows some promise—e.g., [1]). The bottom portion of the figure emphasizes the *signals*, or fixed-length nucleotide motifs, which serve as boundaries for individual exons and introns. Most eukaryotic introns begin with a GT dinucleotide (called a *donor site*) and end with an AG (called an *acceptor site*).

A gene *parse* thus consists of a syntactically valid series of signals from the set $V=\{ATG, GT, AG, TGA, TAA, TAG\}$ which have been identified in the input sequence. The necessary syntactic constraints on the parse of a genomic sequence are:

$$ATG \rightarrow TAG$$
$$ATG \rightarrow GT$$
$$GT \rightarrow AG$$
$$AG \rightarrow GT$$
$$AG \rightarrow TAG$$
$$TAG \rightarrow ATG$$

where the rule $X \rightarrow Y$ indicates that signal X may be followed by signal Y in a syntactically valid parse (rules for genes on the opposite DNA strand are easily obtained from these). The set of all valid parses for a given input sequence may be

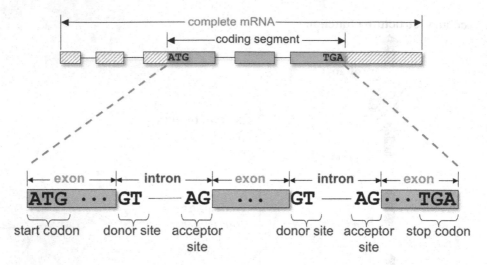

Fig. 2. The gene-parsing problem. A complete mRNA consists of one or more exons (rectangles). Portions of these exons may be coding (gray) or noncoding (hatched), with only the former giving rise to amino acids during translation. The coding segment extends from a start codon (ATG) to a stop codon (TGA, TAG, or TAA), with one or more introns (GT to AG) in between. Introns are spliced out prior to translation into a protein. Source: Majoros WH, *Methods for Computational Gene Prediction*, Cambridge University Press (forthcoming), reproduced with permission.

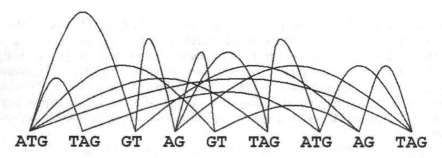

Fig. 3. An example parse graph. Vertices are shown as dinucleotide or trinucleotide motifs at the bottom. Edges denote exons, introns, or intergenic regions. Source: Majoros WH, *Methods for Computational Gene Prediction*, Cambridge University Press (forthcoming), reproduced with permission.

represented using a *parse graph* (Fig. 3) in which vertices represent putative signals and edges represent possible exons, introns, and intergenic regions.

Because not every ATG/GT/AG/TAG/TGA/TAA occurring in a sequence is a true start codon, donor site, acceptor site, or stop codon, as recognized by the living cell, the gene-parsing problem is a highly ambiguous one. For this reason, stochastic parsers based on probabilistic models of DNA have largely come to dominate the gene-finding field. Several of the most popular types of model for this task are described in the sections that follow.

2.2 Hidden Markov Models

A *hidden Markov model* (*HMM*) is a state-based generative model which emits symbols over a finite alphabet. Formally, a hidden Markov model $M = (Q, \alpha, P_t, P_e)$ operates by beginning in the special state $q_0 \in Q$, transitioning stochastically from state to state (i.e., between elements of $Q = \{q_0, q_1, ..., q_{m-1}\}$) according to the *transition distribution*, $P_t(q_j | q_i)$, and emitting a single symbol $c \in \alpha$ according to the *emission distribution*, $P_e(c | q)$, upon entering state $q \in Q$. The machine ceases operation when it re-enters state q_0 (which emits no symbols). Gene-finding with an HMM is accomplished by positing that the DNA sequence S under study was generated by a particular model M having alphabet $\alpha = \{A, C, G, T\}$ and then identifying the most probable *path* (series of states) ϕ^* by which M could have generated S:

$$\phi^* = \underset{\phi}{argmax} \, P(\phi | S) = \underset{\phi}{argmax} \, \frac{P(\phi, S)}{P(S)} = \underset{\phi}{argmax} \, P(\phi, S)$$
$$= \underset{\phi}{argmax} \, P(S | \phi) P(\phi). \tag{1}$$

That is, were the HMM to emit sequence S, the most probable way for it to do so would be for it to pass through precisely the series of states specified by ϕ^*. Eq. (1) can be further factored into a product of emission and transition probabilities along a prospective path ϕ by decomposing $P(\phi)$ into P_t terms, and $P(S | \phi)$ into P_e terms:

$$\phi^* = \underset{\phi}{argmax} \, P_t(q_0 | y_{|S|}) \prod_{i=0}^{|S|-1} P_e(x_i | y_{i+1}) P_t(y_{i+1} | y_i), \tag{2}$$

where $S = x_0 ... x_{|S|-1}$ is a sequence of length $|S|$, for nucleotides x_i, and $\phi = (y_0, ..., y_{|S|+1})$ for states $y_i \in Q$; $y_0 = q_0$ and $y_{|S|+1} = q_0$ are assumed since the machine must begin and end in state q_0. Actually finding the optimal path (or "parse") ϕ^* can be carried out using Viterbi's dynamic programming algorithm [2], which entails the computation of two matrices, $V(i, k)$ for the path probabilities and $T(i, k)$ for the *traceback pointers* which allow us to reconstruct the optimal path once the matrices have been computed:

$$V(i, k) = \begin{cases} \underset{j}{max} \, V(j, k-1) P_t(q_i | q_j) P_e(x_k | q_i) & \text{if } k > 0, \\ P_t(q_i | q_0) P_e(x_0 | q_i) & \text{if } k = 0. \end{cases} \tag{3}$$

$$T(i, k) = \begin{cases} \underset{j}{arg \, max} \, V(j, k-1) P_t(q_i | q_j) P_e(x_k | q_i) & \text{if } k > 0, \\ 0 & \text{if } k = 0. \end{cases} \tag{4}$$

Reconstruction of the optimal path proceeds by starting at the highest-scoring cell (i, k) in the last column of the V matrix and iteratively assigning $i \leftarrow T(i, k)$ and $k \leftarrow k-1$ until the first column ($k=0$) is reached; the successive i visited during this traversal correspond to the states q_i in the optimal path (in reverse order).

Training of an HMM is most commonly carried out using maximum likelihood estimation (MLE). In the simplest case, in which individual nucleotides in the training sequences are labeled with corresponding states in the model, MLE can be performed simply by tabulating the number of times $C(q_i, q_j)$ that state q_i was followed by q_j in the training set, and also the number of times $C(s_k, q_i)$ that nucleotide s_k was labeled with state q_i, for alphabet $\alpha = \{s_k | 0 \ k < size(\alpha)\}$. Normalizing these counts produces the desired probability estimates:

$$P(q_j | q_i) \approx \frac{C(q_i, q_j)}{\sum_{h=0}^{|\mathcal{Q}|-1} C(q_i, q_h)}, \qquad P_e(s_k | q_i) \approx \frac{C(s_k, q_i)}{\sum_{h=0}^{|\alpha|-1} C(s_h, q_i)}. \qquad (5)$$

More sophisticated methods such as *Viterbi training* or the use of an *expectation maximization* (*EM*) algorithm [3] are required when labeled training data are not available [4].

A simple HMM for gene finding is depicted in Fig. 4. The state labeled (N) represents intergenic regions. The machine may self-transition any number of times while in this state to generate arbitrarily long intergenic regions. Following the path $q_2 \rightarrow q_3 \rightarrow q_4$ produces a start codon (ATG) and places the machine in the exon states (q_5, q_6, q_7—three states to represent the three codon positions). Generation of an intron begins with a donor site (GT; $q_{13} \rightarrow q_{14}$) followed by an arbitrarily long intronic region ("I", q_{15}) and then an acceptor site (AG; $q_{16} \rightarrow q_{17}$). The reader can easily verify that states $\{q_8, q_9, q_{10}, q_{11}, q_{12}\}$ generate only the three eukaryotic stop codons, TGA, TAA, and TAG. Note that states labeled with a specific nucleotide in the figure can generate only that symbol (e.g., T for state q_{14}).

Such a simple HMM can be extended in various ways to improve gene-finding accuracy, primarily through the more detailed modeling of statistical biases in nucleotide composition within gene features. An example is the use of higher-order emission probabilities:

$$P_e(x_i | x_{i-n} x_{i-n+1} ... x_{i-1}, q_j) \approx \frac{C(x_{i-n} x_{i-n+1} ... x_i, q_j)}{\sum_{s \in \alpha} C(x_{i-n} x_{i-n+1} ... x_{i-1} s, q_j)}, \qquad (6)$$

where $P_e(x_i | x_{i-n} x_{i-n+1} ... x_{i-1}, q_j)$ denotes the probability of state q_j emitting symbol x_i, given that the subsequence $x_{i-n} ... x_{i-1}$ has just been emitted; counts $C(x_{i-n} x_{i-n+1} ... x_i, q_j)$ for all ($n+1$)-letter sequences $x_{i-n} x_{i-n+1} ... x_i$ may be derived from the training data as before.

An unfortunate aspect of gene modeling with HMMs is the fact that variable-length features (such as exons or introns) are implicitly modeled as having geometrically

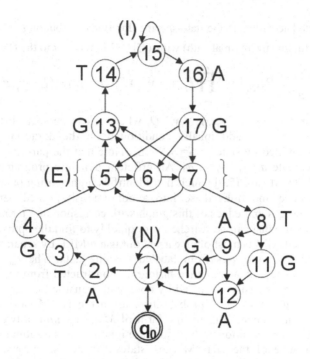

Fig. 4. A simple HMM for gene finding. States are represented as circles and transitions as arrows. Probabilities are omitted for clarity. States which emit only one symbol are shown with the corresponding symbol next to the state. The special state q_0 is the start/stop state, which emits no symbols. Source: Majoros WH, *Methods for Computational Gene Prediction*, Cambridge Univer sity Press (forthcoming), reproduced with permission.

distributed lengths, as enforced via the compounding of repeated transition probabilities during generation of a variable-length feature. Generalized HMMs (*GHMMs*—see below) solve this problem while also allowing for greater modeling flexibility.

2.3 Generalized Hidden Markov Models

GHMMs improve on HMMs by abstracting the generation of entire gene features into single states; i.e., upon entering a state q_i the machine may emit an entire subsequence S_i before making the next transition. In this way, feature lengths may be explicitly modeled via arbitrary distributions (not necessarily geometric), the syntactic and statistical properties of individual features may be encapsulated within each state in an arbitrary (i.e., non-Markovian) way, and the number of states required to implement a production-quality gene-finding system can be kept relatively small.

Formally, a GHMM is a stochastic generative model $M = (Q, \alpha, P_t, P_e, P_d)$ in which all terms are as defined for the HMM case, except that individual state emissions are entire substrings (rather than individual symbols) over α, with those emissions having lengths distributed according to the state-specific duration distribution $P_d(L|q)$, $L \in \mathbb{N}$, $q \in Q$. Decoding (i.e., finding the optimal path) with a GHMM is similar to the HMM case:

$$\phi^* = \underset{\phi}{argmax} \, P_t(q_0 \mid y_n) \prod_{i=1}^{n} P_t(y_i \mid y_{i-1}) P_d(d_i \mid y_i) P_e(S_i \mid y_i, d_i) \,, \qquad (7)$$

for putative parse $\phi = (y_0, \dots, y_{n+1})$, $\forall_i y_i \in Q$, where it can be seen that the emission term $P_e(S_i | y_i, d_i)$ is now additionally conditioned on the duration $d_i = |S_i|$ of the subsequence S_i emitted by state y_i; $S = S_1 S_2 \dots S_n$. Note that the parse again begins and ends in (silent) state q_0: $y_0 = y_{n+1} = q_0$. An efficient dynamic programming heuristic exists for the GHMM case [5,6] which first identifies high-scoring putative signals in the input sequence and links these into a continuously-pruned parse graph; by weighting the vertices and edges of this graph with corresponding terms from Eq. (7) we obtain a structure that can be searched very quickly to find the optimal parse.

Training of a GHMM is most often carried out using MLE by separately estimating the P_t, P_e, and P_d parameters from labeled training data. The P_d distribution is commonly represented via a smoothed histogram constructed from feature lengths in the training data; P_t is easily estimated by observing transition counts in the training data and normalizing these into probabilities, as in the HMM case. Because most GHMM-based gene finders utilize some form of *Markov chain* (a two-state, higher-order HMM in which transition probabilities are ignored) as the submodel within each variable-length state of the GHMM (i.e., states for exons, introns, or intergenic regions), estimation of P_e is rendered trivial; interpolation techniques are also sometimes employed to mitigate the effects of sampling error when using higher-order models [7]. Fixed-length states of the GHMM, which correspond to signals such as start/stop codons and donor/acceptor sites, are typically represented using a *weight matrix* (*WMM*) [8], in which each position of a fixed-length signal window is described by a position-specific emission distribution, possibly conditional on the symbols residing at other positions within the window [9]. Thus, for most GHMM implementations, MLE parameter estimation may be performed without the need for iterative methods such as Viterbi training or EM.

2.4 Pair HMMs and Generalized Pair HMMs

A significant increase in predictive accuracy can often be achieved by modeling evolutionary trends as observed in the genomic sequence of related organisms. This is due to the fact that natural selection tends to operate more stringently on the coding (versus noncoding) regions of any genome. When predicting genes in some target genome S, an informant genome I from some related organism may be employed by aligning portions of S and I for which *homology* (evolutionary commonality of descent) may be inferred via sequence similarity. In this case, the optimal parse may be defined as that ϕ which maximizes $P(\phi|S,I)$, which we may factor as:

$$\phi^* = \underset{\phi}{arg\,max} \, P(\phi \mid S, I) = \underset{\phi}{arg\,max} \, \frac{P(\phi, S, I)}{P(S, I)} = \underset{\phi}{arg\,max} \, P(\phi, S, I)$$

$$= \underset{\phi}{arg\,max} \, P(\phi) P(S, I \mid \phi), \qquad (8)$$

where $P(\phi)$ is merely the product of transition probabilities incurred along the path ϕ just as before, leaving only the problem of evaluating $P(S,I|\phi)$. A particularly elegant method for modeling the latter joint probability is by positing a special type of Markov model $M=(Q, \alpha, P_t, P_e)$ in which each state $q \in Q$ emits pairs of symbols $(s_1, s_2) \in \alpha \times \alpha$ rather than individual symbols as in a standard HMM. Replacing α with the augmented alphabet $\alpha^- = \{A,C,G,T,-\}$, for '$-$' the gap symbol representing unaligned positions or gaps in an alignment, we arrive at a model capable of emitting aligned sequences with gaps. Such a model is called a *Pair HMM* (*PHMM*); an example is shown in Fig. 5.

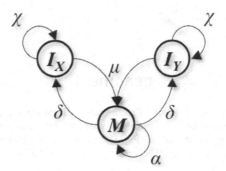

Fig. 5. A simple Pair HMM. State M emits matched or mismatched symbols into an alignment; I_X and I_Y emit gapped alignment positions (i.e., gaps in sequence X for I_X and in sequence Y for I_Y). Transition probabilities are indicated using letters α, χ, δ, and μ. Source: Majoros WH, *Methods for Computational Gene Prediction*, Cambridge University Press (forthcoming), reproduced with permission.

Decoding with a PHMM may be described as:

$$\phi^* = \underset{\phi=\{y_0,...,y_{n+1}\}}{\arg\max} P_t(q_0 \,|\, y_n) \prod_{i=1}^{n} P_e(a_{i,1}, a_{i,2} \,|\, y_i) P_t(y_i \,|\, y_{i-1}) , \qquad (9)$$

where $a_{i,j}$ denotes the ith symbol in the jth track ($j \in \{1,2\}$) of the alignment formed by a putative parse ϕ. Unfortunately, a dynamic programming solution to this optimization problem requires a three-dimensional matrix and therefore significantly greater computational resources than for a standard HMM. Heuristics are thus commonly employed to prune the matrix, as illustrated in Fig. 6. The heuristic aligner *BLAST* [10] is often used to precompute a set of *guide alignments* (black bars in the figure); portions of the dynamic programming matrix which are deemed too distant from these guide alignments are pruned from the matrix and never evaluated.

Generalized PHMMs (GPHMMs) have also been employed for gene prediction [11,12]. A GPHMM may be obtained by embedding a PHMM within each state of a GHMM, so that each GPHMM state will emit pairs of aligned genomic features (e.g.,

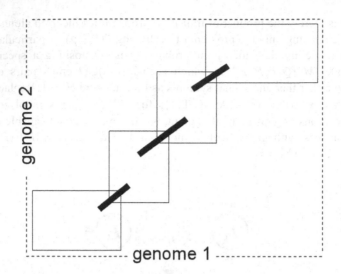

Fig. 6. Pruning an alignment matrix. Precomputed alignments are shown as solid bars; rectangles denote the portion of the alignment matrix which are actually evaluated. The third dimension, corresponding to states of the PHMM, is omitted for clarity. Source: Majoros WH, *Methods for Computational Gene Prediction*, Cambridge University Press (forthcoming), reproduced with permission.

exons, introns, or intergenic regions). The corresponding decoding optimization is given by:

$$\phi^* = P_t(q_0 \mid y_n) \overset{\arg\max}{\underset{\phi}{}} \prod_{i=1}^{n} P_e(S_{i,1}, S_{i,2} \mid y_i, d_{i,1}, d_{i,2}) \tag{10}$$

$$P_t(y_i \mid y_{i-1}) P_d(d_{i,1}, d_{i,2} \mid y_i).$$

One dynamic programming solution for GPHMMs proceeds by constructing a parse graph (Fig. 3) for each of the two input sequences and then aligning these graphs in such a way that like vertices (e.g., ATG-ATG, GT-GT, etc.) are permitted to align and unlike vertices (e.g., ATG-TAG) are not, with Eq. (10) serving as the objective function of the alignment process [12]. The resulting alignment between parse graphs will outline an isomorphism corresponding to a parse in each of the two graphs. Precomputed guide alignments are generally also required for GPHMMs in order to achieve acceptable time-space complexity via pruning of the dynamic-programming matrix.

2.5 Phylogenetic HMMs

Whereas PHMMs and GPHMMs incorporate homology evidence from a single informant genome, *Phylogenetic HMMs* (*PhyloHMMs*) permit evidence from any number of informants to be utilized, with a Bayesian network being employed to reduce bias due to the non-independence of the informants. Precomputed alignments are again used; unlike PHMMs and GPHMMs, however, current PhyloHMM implementations adhere strictly to the precomputed alignments, rather than merely using them as guides for the purpose of pruning the search space.

The decoding derivation for a PhyloHMM is:

$$\phi^* = \underset{\phi}{\arg\max} \; P(\phi \mid S, I^{(1)}, ..., I^{(n)}) \tag{11}$$

$$= \underset{\phi}{\arg\max} \; \frac{P(\phi, S, I^{(1)}, ..., I^{(n)})}{P(S, I^{(1)}, ..., I^{(n)})}$$

$$= \underset{\phi}{\arg\max} \; P(\phi, S, I^{(1)}, ..., I^{(n)})$$

$$= \underset{\phi}{\arg\max} \; P(\phi) P(S, I^{(1)}, ..., I^{(n)} \mid \phi)$$

$$= \underset{\phi}{\arg\max} \; P(\phi) P(S \mid \phi) P(I^{(1)}, ..., I^{(n)} \mid S, \phi),$$

for target genome S and informants $I^{(1)}, ..., I^{(n)}$. The $P(\phi)P(S \mid \phi)$ term can be evaluated using a standard GHMM decoder. The remaining term, $P(I^{(1)}...I^{(n)} \mid S, \phi)$, can be evaluated as follows:

$$P(I^{(1)}, ..., I^{(n)} \mid S, \phi) = \prod_{y_i \in \phi} \prod_{j=b_i}^{e_i} F(I_j^{(1)}, ..., I_j^{(n)} \mid S_j, \psi_i), \tag{12}$$

where the second product is over columns b_i through e_i of the precomputed alignment, according to the emission of state $y_i \in \phi$. The evolution model ψ_i typically differs between coding states (i.e., ψ_{coding}) and non-coding states ($\psi_{noncoding}$) so as to model the differences in rates of evolution between the coding and noncoding portions of genomes. These rates are reflected in the $F(\cdot)$ term, which is known as *Felsenstein's algorithm* [13], and is used to compute the likelihood of a single column in the alignment:

$$F(I_j^{(1)}, ..., I_j^{(n)} \mid S_j, \psi_i) = \sum_{unobservables} \left(\prod_{\substack{nonroot \\ v}} P(v_j \mid parent(v_j), \psi_i) \right), \tag{13}$$

where the summation is over all possible assignments of nucleotide sequences to the (unobserved) ancestral species in a *phylogenetic tree* (or *phylogeny*) describing the evolutionary relationships among the target and informant genomes; the phylogeny effectively serves as a Bayesian network for modeling evolutionary dependencies. v_j is the residue in column j of the alignment for any non-root vertex v in the tree; these v_j thus correspond to the (observable) informants as well as their (unobservable) common ancestors in the phylogeny. Summing over all the possible nucleotides in the ancestral genomes permits us to evaluate this formula in the presence of unobservables, by effectively computing an expectation. Before this computation may be performed the phylogeny must first be re-rooted so that the target genome is at the root of the phylogeny and the informant genomes are at the leaves [14], reflecting the dependence of the informants on the target.

The actual dependence of each genome on its parent genome—denoted $P(v_j|parent(v_j), \psi_i)$ in Eq. (13)—may be represented at the individual nucleotide level using a substitution matrix $\mathbf{M}$ in which the entry $\mathbf{M}_{a,b}$ gives the probability of nucleotide a evolving into nucleotide b during a period of time equivalent to the evolutionary distance between the two genomes. Non-independence of the columns in the alignment may be modeled as well, by conditioning the substitution matrix on one or more preceding nucleotides in the parent genome, similar to the higher-order Markov models described earlier.

The substitution matrices comprising the evolutionary models ψ_i of a PhyloHMM may be independently trained from aligned features of the appropriate type (e.g., aligned coding exons for ψ_{coding}) using standard maximum likelihood techniques developed prevously for phylogeny reconstruction [15]. A general-purpose gradient ascent procedure may thus be employed to maximize the likelihood of the training data using Eq. (12) as the objective function of the optimizer.

2.6 Ad Hoc "Combiner" Methods

Integration of other forms of evidence besides evolutionary conservation between genomes—such as expression evidence in the form of messenger RNAs and proteins culled from living cells of the target organism—can be incorporated as well, though current methods tend to be largely *ad hoc* in nature and therefore defy (at present) any concise, unified description such as those given in the preceding sections. These programs are referred to as *combiners*, since they may combine many disparate sources of evidence, including predictions from other gene-finding programs. Despite their typically *ad hoc* nature, some combiner programs have proven to be among the most accurate systems currently available for predicting gene structure [16,17]. It seems a curious fact that, despite their not conforming (in most cases) to a rigorous probabilistic formulation as in the case of Markov models and their various relatives described earlier, combiner-type programs can perform so well. As we will discuss in greater detail below, this may be due (in part) to the fact that combiner systems are typically trained *discriminatively* via extensive manual tuning of evidence weights, with the goal of the manual tuner being to maximize the accuracy of the gene predictions when the system is applied to the sequences in the training set, as opposed to maximizing the *likelihood* of the training data as in MLE. Another likely reason for the success of combiners is their integration of all available forms of evidence in arriving at a prediction; it is in reference to this latter property that combiner-type programs are often referred to as being *integrative*. Unfortunately, because the "gold standard" against which gene finders are often measured—namely, test sets of previously annotated genes—is often produced (or at least heavily influenced) by a combiner-like "annotation pipeline" (see section 4.5), the superiority of integrative systems may in fact be somewhat over-estimated.

3 Limitations of Current Methods

3.1 MLE+Viterbi Is Not Optimal

As described above, most state-of-the-art gene-finding systems are at present based on Markovian models of one type or another (i.e., HMMs, GHMMs, PHMMs, GPHMMs, PhyloHMMs). The vast majority of systems based on these models are trained via MLE and are then subjected to some form of Viterbi decoding, with the latter being extended in various ways to incorporate external evidence such as informant sequences (e.g., PHMMs and PhyloHMMs) as well as modeling enhancements such as explicit state duration (e.g., GHMMs). Much evidence suggests, however, that these MLE-trained systems are not optimal in practice, in that the use of non-maximum-likelihood parameters can often improve the accuracy of a given probabilistic parser when the parser is later utilized for Viterbi-based prediction. Indeed, the suboptimality of the MLE+Viterbi strategy has been well-documented for some time now in the field of speech recognition, in which HMM-based systems are fairly routinely subjected to one of several non-MLE forms of training collectively known as *discriminative training* [18-20].

Whereas the goal of maximum likelihood training is to maximize the joint likelihood of the training set T (consisting of pairs of sequences S and their "correct" parse ϕ) given the model parameters θ—e.g.,

$$\theta^*_{MLE} = \underset{\theta}{\arg\max} \left(\prod_{(S,\phi) \in T} P(\phi, S \mid \theta) \right), \tag{14}$$

the goal of discriminative training is to maximize the *expected accuracy* of the resulting parser. This latter goal can be formalized in a number of ways. A common formulation is the so-called *conditional maximum likelihood* (*CML*):

$$\theta^*_{CML} = \underset{\theta}{\arg\max} \left(\prod_{(S,\phi) \in T} P(\phi \mid S, \theta) \right) = \underset{\theta}{\arg\max} \left(\prod_{(S,\phi) \in T} \frac{P(\phi, S \mid \theta)}{P(S \mid \theta)} \right), \tag{15}$$

in which we require the parameterization θ^*_{CML} under which the correct parses of the training sequences are most probable, given the sequences and the model parameters. Unfortunately, methods for directly optimizing Eq. (15) for an HMM are not known [4], and while a number of heuristics have been developed within the field of speech recognition for this or similar objective functions (e.g., *maximum mutual information, MMI* [18]; *minimum classification error, MCE* [19]), these tend to be unstable in practice so that convergence is typically not guaranteed without manual tuning of additional parameters (e.g., [19, 21]). It should also be noted that for practical gene finders the number of model parameters to be optimized can be in the high thousands in the case of higher-order models, making thorough discriminative training of such models seem highly daunting at best.

Explicit discriminative training for HMM-based gene finders has thus been largely ignored (see [21] for a rare example). In the case of GHMMs and more sophisticated probabilistic models for gene finding, much anecdotal evidence suggests that a very crude form of discriminative training is typically performed via manual tuning of a small number of model parameters by the authors of these systems so as to improve the observed prediction accuracy on the training set or on a separate test set. In the case of *comparative* gene-finding systems (i.e., those incorporating external evidence apart from the target genome), such manual tuning is commonly performed by introducing one or more "fudge factors" to allow for the artificial weighting of the various components of the decoding objective function such as the informant component (e.g., "coding bias" in *ExoniPhy* [15]; "conservation score coefficient" in *N-SCAN* [22]; non-maximum-likelihood value for P_{match} in *TWAIN* [12]). Though these "fudge factors" appear to serve no theoretical role in the probabilistic formulation of the model, such manipulations can sometimes dramatically increase the accuracy of the resulting parser.

Automated discriminative training procedures for generalized HMMs and comparative systems such as pair HMMs and PhyloHMMs have received little or no attention as of yet. A rare exception involved the use of a crude gradient-ascent approach to optimize a handful of the thousands of parameters making up a GHMM-based gene finder [23]. Given the simplistic and *ad hoc* nature of the "fudge factor" approach described above for PhyloHMMs and other sophisticated probabilistic gene parsers, investigations into more comprehensive means of discriminatively optimizing these systems would seem to be well justified.

Alternatively, one might consider the very need for discriminative training of Markovian gene-finding models to be an indication that this family of models is perhaps not an ideal one for the gene-finding application. Investigations into explicitly discriminative, non-Markovian frameworks such as *conditional random fields* have recently produced promising preliminary results [24, 25]. The use of alternate HMM decoders (i.e., in place of Viterbi) remains another possibility, though experiments by ourselves with two recently-proposed alternate decoders (*posterior Viterbi* [26], *optimal accuracy decoder* [27]) suggest that these decoders do not provide an appreciable gain in predictive accuracy for eukaryotic gene finding, and in particular do not obviate the need for discriminative training of the model (unpublished data).

3.2   Reliance on Precomputed Alignments

As mentioned earlier, the PhyloHMM framework, and to a lesser extent the PHMM and GPHMM frameworks, rely on pre-computed alignments of the target and informant genomes to be used during gene prediction. In the case of Pair HMMs and GPHMMs, the pre-computed alignments serve largely as guides, so that the actual pairing off of target and informant nucleotides resulting from a decoding run of the system may differ to some degree from that prescribed by the pre-computed alignment, though in practice the aggressive pruning of the dynamic programming matrix around the guide alignments may preclude all but the smallest divergence from the pre-computed alignment. In the case of PhyloHMMs, all known implementations at present adhere to the pre-computed alignment precisely, so that alignment errors by the external alignment tool may give rise to spurious evolutionary patterns as seen by the PhyloHMM decoder. Ideally, one would like the gene prediction and alignment phases to proceed simultaneously, so as to mutually inform one another, as in the case of (non-pruned) PHMM decoding. Methods for efficiently achieving this in the case of PhyloHMMs have yet to be investigated.

3.3 Simplifying Assumptions

A number of simplifying assumptions are typically made in formulating a gene-finding model, most often for the purpose of reducing the computational complexity of the decoding process. In particular, various models assume that:

1. feature lengths are geometrically distributed (HMMs)
2. exon-intron structure does not change over evolutionary time (GPHMMs, PhyloHMMs)
3. pre-computed alignments are correct (PhyloHMMs; also to some degree GPHMMs and PHMMs)
4. each locus has exactly one correct parse (one "isoform")
5. the target sequence contains no frameshifts
6. genes do not overlap
7. non-consensus splice sites do not occur
8. stop codons do not code for any amino acid

Though all of these assumptions can be shown to be false in at least one biologically valid instance, few efforts have been undertaken to relax these assumptions. Known exceptions include the modeling of non-geometrically distributed intron lengths [28] and the modeling of genes which overlap on opposite strands [29], neither of which have seen widespread adoption in mainstream eukaryotic gene finders as of yet. In the case of non-consensus splice sites, though several software implementations do permit the user to explicitly request the modeling of non-consensus splice sites, a thorough analysis of the impact of this feature on prediction accuracy has yet to be performed, while conventional wisdom holds that the sensitivity gains can be more than offset by the loss in specificity.

Because Markovian-based gene finders utilize a Viterbi decoding step to find the single most promising parse of an input sequence, any genes which are predicted as part of the parse will be assigned a single exon-intron stucture by the gene finder. Unfortunately, many human genes (perhaps as many as 80%) can be spliced in multiple ways to produce distinct intron-exon structures, or *isoforms*. The issue of multiple isoforms is discussed in more depth in the next section.

The assumption that stop codons do not code for any amino acid is untrue in the very rare case of *selenocysteine*—an amino acid coded by the codon TGA (UGA in the mRNA). In general, gene finders do not predict genes containing in-frame stop codons (i.e., stop codons residing at a distance d from the beginning of the coding portion of the spliced gene, in which d is divisible by 3), except for the in-frame stop codon occurring at the very end of the gene. For most organisms, to allow the prediction of genes with in-frame stop codons (other than the termination codon at the end of the gene) would very likely result in a significant degradation in predictive accuracy, since for most sequenced genomes to date, the majority of known genes do not contain in-frame stop codons. A rare example of a gene-finding system which can predict selenocysteine-bearing genes has been described [30] in which homology evidence and other information from the UTR of a putative gene were used to limit the large number of possible in-frame stop-codon-bearing genes to a more reasonable number.

The assumption that genes do not overlap is specific to eukaryotic gene finders; because overlapping genes appear to be more common in prokaryotes, prokaryotic gene-finding programs have modeled overlapping genes for some time now [31, 32] and gene finders for eukaryotic viruses such as HIV also must deal with the phenomenon of overlapping genes [33]. In the case of eukaryotes, nested genes and genes which overlap other genes on the opposite strand are not just rare exceptions (e.g., in *Drosophila melanogastor* [34]), though most eukaryotic gene finders do not predict them. Two exceptions are SNAP [29] and AUGUSTUS [28], which can be run in a special single-strand mode, in which genes are independently predicted on either strand, so that a gene prediction on one strand may overlap a prediction the other strand.

Reliance on pre-computed alignments has already been discussed; the somewhat related issue of conservation of exon-intron structure in GPHMMs and PhyloHMMs is similarly vexing. Fig. 7 illustrates the problem for a pair of *Aspergillus* homologues. The upper track in the figure depicts the exon-intron structure of a particular gene in *A. oryzae*; the lower track depicts the homologous gene in *A. fumigatus*, where it can be seen that a number of structural changes have been effected since these organisms diverged from their common ancestor, though the encoded proteins have remained identical. Efficient GPHMM implementations generally do not permit the prediction of homologues with different exon-intron structures, since to do so would largely eliminate any opportunity for pruning the search space, resulting in dynamic programming matrices which are often too large to evaluate in a reasonable amount of time. In the case of PhyloHMMs, the potential for such structural changes would at the least seem to present a challenge for the alignment pre-processing phase. More specifically, the need for incorporating amino

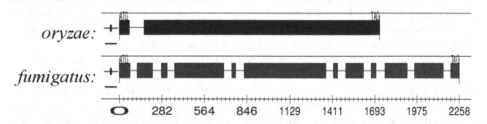

Fig. 7. An example of exon-intron structure divergence. These two genes from *Aspergillus oryzae* and *A. fumigatus* encode the same protein, but have accumulated a number of structural changes since their last common ancestor. Many comparative gene finders cannot easily model such structural changes. Source: Majoros WH, *Methods for Computational Gene Prediction*, Cambridge University Press (forthcoming), reproduced with permission.

acid conservation into the alignment phase would seem to be greater than is perhaps recognized at present.

3.4 The Existence of Alternative Splicing

The propensity for human genes to encode multiple, distinct proteins via alternative splicing (as well as alternative polyadenylation and alternative transcription/translation initiation) is now well documented [35]; Fig. 8 illustrates some of the potential effects of alternative splicing and related phenomena.

Each potential splicing pattern gives rise to a unique *isoform* for the locus. Some loci can have very many isoforms [36], and there is even evidence that exons from distinct loci in the human genome may sometimes be spliced together to encode a "chimeric" protein [37]. It has been suggested that the propensity for a locus to encode multiple proteins may account for the seemingly large mismatch between the estimated number of human genes (~25000) and the number of proteins (>100000), and is therefore a particularly important issue for human gene finding.

Despite the prevalence of these phenomena in human genes, however, virtually all state-of-the-art eukaryotic gene finders continue to enforce a one-gene-one-parse discipline via their use of Viterbi (or Viterbi-like) decoding to find the single optimal parse of the input sequence. We will address possible methods for relaxing this discipline in section 4.1.

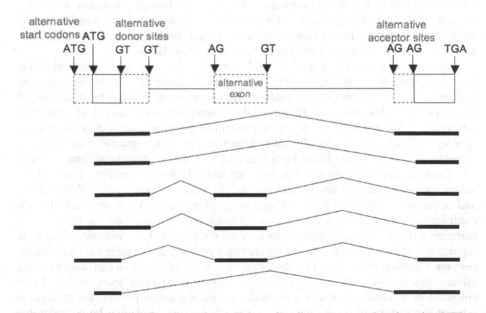

Fig. 8. Some possibilities for alternative splicing of coding segments (i.e., ignoring UTRs). Many isoforms may potentially be produced from a single locus in a combinatorial fashion. Source: Majoros WH, *Methods for Computational Gene Prediction*, Cambridge University Press (forthcoming), reproduced with permission.

4 Some Possible Future Directions

4.1 Redefining the Problem

The earliest "gene finding" systems were actually exon finders: that is, rather than predicting complete gene structures, they instead predicted individual exons, and left the task of assembling exons into complete genes to the end user. As Markov-based systems gained in popularity it became more feasible to predict whole gene structures via the well-established Viterbi decoding algorithm. As the prevalence of alternative splicing in mammalian genomes becomes better appreciated, however, the suitability of a Viterbi-based approach is increasingly cast into doubt. A modified version of Viterbi decoding which permits the efficient identification of the N best (rather than the single best) parses has been suggested as one possible means of addressing the issue of alternative splicing within current gene-finding frameworks [38]. However, not all possible valid alternative isoforms are actually produced in an organism, and without additional splicing-specific information, we will not be able to deduce the set of isoforms which are actually produced.

One possible remedy lies in redefining the problem so as to focus on the identification of likely exons in isolation—i.e., predicting individual exons without regard to their compatibility (i.e., whether they overlap, whether they maintain a consistent reading frame, etc.) with other predicted exons in a complete gene parse. The task of assembling these exon predictions into one or more predicted isoforms for a locus can then be left for downstream software, or for human annotators in the case of well-funded genome projects. Although this redefinition of the problem would seem to be a step backward toward the earlier exon-finding approaches mentioned above, there are a number of potential advantages to this change.

The most obvious advantage of such an approach, for organisms exhibiting appreciable levels of alternative splicing, is that it facilitates the identification of multiple isoforms by downstream analyses after exon prediction has been performed. For instance, the last few years have seen the development of a number of algorithms which allow the predictions of which individual exons are subject to alternative splicing [39-41], and additionally other alternative splicing patterns such as *intron retention* [42]. That is, the identification of likely exons and the assembling of exons into multiple isoforms become effectively decoupled, thereby entailing many of the advantages of modular software design (i.e., division of labor, ease of development and debugging, efficiency gains through parallelization, etc.). Given a set of high-confidence exon predictions from an exon finder, research into optimal methods for combining these into multiple-isoform predictions may proceed without the need to repeatedly perform the content-scoring analyses encapsulated within the exon finder, perhaps significantly easing the computational load of development and research efforts. Indeed, were the exon predictions from one or more exon finders to be collected into publicly available data banks for each genome project, the annotation (and re-annotation) of these genomes at the whole-gene level may be considerably eased, since the exon-finding phase need not be performed anew as alternative parameterizations of the exon-assembly process are explored. Exons predicted by different exon finders may also be considered for combination by automated methods into coherent isoform predictions (thereby addressing the not-uncommon situation in which one gene finder correctly predicts one exon of a gene while another gene finder correctly predicts another, but neither program predicts the entire gene correctly).

Predicting individual exons for later use by an exon-assembly process poses the question of how best to settle the tradeoff between sensitivity and specificity. Many, if not most, exon-finding approaches require that the user or designer impose a scoring threshold below which a putative exon is not reported. In situations in which a later automated exon-assembly process is to be performed, a reasonably liberal threshold would presumably be of greatest value, so as to avoid limiting sensitivity. In a similar vein, one might view an *ensemble* of exon predictions much like a "particle cloud" in statistical physics, in which a particle's position is not precisely defined, but is instead characterized by a probability distribution. In a similar way, one or more exon finders may be used to induce a probability distribution on the set of all possible *open reading frames* (i.e., possible coding exons) in a sequence. To the extent that an exon finder cannot identify exact exon boundaries with absolute certainty (e.g., in cases of alternative splicing affecting the choice of either 5' or 3' splice site), some form of "exon cloud" representation may be appropriate so as not to unduly constrain a downstream exon-assembly process. Because optimal exon assembly in the case of genes with multiple isoforms is not yet a solved problem, such an ensemble-based approach to exon prediction may indeed be a promising starting point. As our knowledge about splicing regulatory factors and their cis-regulatory sequences increases (see, e.g., [43]), we can use information about, e.g., their expression values as evidence to infer condition-specific isoforms.

4.2 A Greater Role for Machine Learning

The redefinition of the gene-finding problem via the decoupling of exon finding from the later assembly of exons into one or more isoforms for each putative gene would in some ways seem to permit a greater role for alternative machine-learning approaches in the gene prediction process. Although a number of machine learning methods have been utilized within gene finders in the past (e.g., decision trees in *GlimmerM* [44]; neural networks in *GRAIL* [45]), the newest generation of gene-finding systems are based primarily on Markov models and generally do not incorporate any other machine learning algorithms. One obstacle to the greater utilization of other machine learning methods in gene finding appears to be the fundamental mismatch between the *classification-oriented* formulation of many machine-learning algorithms (at least the more popular ones such as support vector machines and the like) and the *parsing-oriented* interface of HMMs provided by Viterbi decoding. Because alternative splicing was for a number of years considered a rare exception to the one-gene-one-protein "rule," the single-parse approach enforced by Viterbi decoding became well entrenched in the gene-finding field. Exon finding, on the other hand, permits a very natural interpretation within the classification framework: given an open reading frame, an exon finder aims to accurately classify the interval as being an exon (class 1) or not being an exon (class −1).

Reformulating the problem as one of classification would permit designers of exon-finding software to draw more fully on the vast body of research from the machine-learning field. In particular, the use of maximum discrimination classifiers may produce appreciable accuracy gains as compared to the standard MLE-trained

Markov models which currently dominate the field. This in turn highlights yet another advantage of a move away from the MLE+Viterbi strategy for whole-gene prediction, which as we noted earlier can be characterized as sub-optimal in certain regards.

A particularly popular machine-learning method, *support vector machines* (*SVMs*) [46], has been applied to the problems of exon prediction [47], start codon prediction [48], splice site prediction [49], and the prediction of specific forms of alternative splicing [39]. The discriminative nature of SVMs and the high accuracy rates which have been observed in a number of applications suggest that further investigations into their use for gene and exon prediction may indeed be worthwhile.

4.3 Focus on Integrative Methods

As we noted earlier, the *ad hoc* methods exemplified by so-called "combiner" systems have proven in some cases to be exceptionally effective at producing highly accurate gene predictions, though it seems obvious that much of the advantage enjoyed by these systems derives not so much from their *ad hoc* nature as from their access to multiple forms of evidence (e.g., homology evidence, known proteins, other gene predictions) in making informed decisions regarding the most likely exonic structure for a gene. Despite the success of integrative approaches utilizing all available evidence, much attention in the field remains focused on systems utilizing only limited forms of evidence—e.g., nucleotide-based conservation in the case of PhyloHMMs and other comparative gene finders. A greater emphasis on the further development of integrative approaches to computational gene prediction may thus be useful, though it is acknowledged that in the case of genomes for which little additional evidence besides the primary genomic sequence is available, the advantage of integrative approaches dwindles.

4.4 Interoperability

Yet another possible avenue for advancing the state of the art in computational gene finding is through the use of explicit graph-based representations of genome content. Recall from section 2.1 our definition of a parse graph as a directed acyclic graph in which individual vertices represent putative splice sites and start/stop codons, and edges denote putative exons, introns, and intergenic regions. While not all gene finders explicitly construct such a graph, it is arguably the case that most, if not all, state-of-the-art whole-gene prediction systems construct such a graph implicitly during their processing of the input sequence. For many of these systems, at the point in their decoding algorithms (whether Viterbi or otherwise) when they select an optimal predecessor signal for linking into the "trellis" which is later used to retrace the optimal parse, if the potential predecessors of the current signal are instead linked to the current signal via a weighted edge (with some function of each predecessor's inductive score serving as the weight), then a parse graph would be automatically induced, and could be emitted by the program in addition to (or even instead of) the gene prediction corresponding to the optimal parse.

Such weighted parse graphs could be immensely useful for later re-processing, especially as additional evidence becomes available which was not present at the time the gene finder was originally run. Parse graphs from multiple gene finders (perhaps based on different training sets or utilizing different classes of model) could conceivably be combined with each other and/or with additional evidence (e.g., homology evidence, expression evidence, etc.) to produce a re-weighted graph that may permit more accurate decoding by virtue of the integrative nature of the graph's construction. Decoding of (i.e., extracting a gene prediction from) parse graphs can be done very simply and efficiently using a specialized shortest-path algorithm entirely anologous to Viterbi decoding [6]. Given a standard file format for the storage of such graphs, decoding of any graph could then be performed by a "universal decoder" program, which need not be aware of the actual methods employed in weighting any particular graph. Given the existence of such a "universal decoder," the implementation of a decoder in any given graph-emitting gene finder then becomes unnecessary, since the universal decoder may be applied to the emitted graph. Were such a graph-based interface to be adopted by a sufficient number of gene-finding systems, entire pipelines may conceivably be constructed in which the graphs from one or more gene finders are subjected to any number of re-weighting processes to incorporate additional information such as the existence of *genomic repeats* [50] or other genome-level features not commonly utilized by the primary gene-finding programs, or which were not available when the programs were trained. The last stage in such a pipeline would presumably involve the use of a graph-based decoder to extract one or more gene predictions.

The utility of a graph-based representation for the identification of alternative splicing should be fairly obvious. Indeed, graph-based methods for the identification of alternative splicing have already been proposed, though not in an overtly Markovian setting [51]. In our own research we have observed a tendency for our graph-based gene finders to often rank the "correct" gene parse very highly, while ranking another, incorrect parse only slightly higher, so that were the program to emit the top N parses, for some reasonably small N, instead of the single highest-scoring parse, the correct parse would very often be among the top N. Because most state-of-the-art eukaryotic gene finders emit only the single highest-scoring parse, the "correct" parse (which might be recognized by a human annotator as correct, due to his or her access to additional evidence) is effectively lost. Methods for sampling parses from an HMM have been explored, and their possible utility to the detection of alternative splicing suggested [38], though the actual adoption of these methods by mainstream gene finders has for the most part not occurred. The proposed practice of emitting an entire parse graph (after applying a reasonable amount of pruning so as to keep the size of the graph manageable while eliminating very unlikely parses) may be viewed as an extreme variant of the sampling approach.

Finally, we would speculate that the availability of pre-computed parse graphs for a large number of organisms in some publicly-available repository—much like the precomputed whole-genome alignments maintained at such sites as the UCSC [52]—may prove useful in enabling researchers to re-analyze genomes at a later date when additional evidence becomes available, without having to deal with the often vexing problem of re-aquiring an older gene finder which had been used in an earlier analysis, or even having to recompile old, possibly poorly-maintained source code in order to run such programs on newer assemblies of a previously annotated genome.

Yet other advantages to graph-based gene prediction conceivably exist which we have not here enumerated. Unless and until a sufficient number of gene-finding software systems adopt such an interface, these advantages will of course prove elusive.

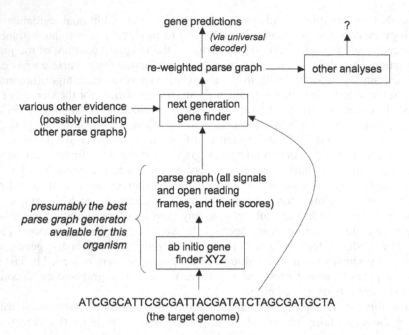

Fig. 9. Some possible uses of parse graphs as a data interchange format for computational gene prediction. Graphs produced via one gene finder may be re-weighted by other downstream programs through the incorporation of additional evidence. Eventually a graph may be supplied to a "universal decoder" to extract an optimal parse. Source: Majoros WH, *Methods for Computational Gene Prediction*, Cambridge University Press (forthcoming), reproduced with permission.

4.5 Improved Evaluation Protocols

It is an unfortunate (and often quite vexing) fact that the unbiased evaluation of gene-finder accuracy can often be very difficult to achieve. Sustained progress in any field depends to a significant degree on our ability to accurately measure progress when it is made. In the case of gene prediction, verification of predicted genes in the laboratory can be rather expensive, so that accuracy assessments are most often made by applying a new (or newly retrained or modified) program to a "test set" of genes for which the intron-exon structures are more-or-less known. Unfortunately, many genes for which we believe we know the "correct" intron-exon structure may in fact be alternatively spliced, so that the predictions obtained for a particular locus which do not agree with the known structure of the gene may in fact match a valid, but unknown, isoform for that gene. In other cases, the "known" structure of a test gene may in fact derive from an earlier gene prediction which had been elevated to "known gene" status by an over-eager human annotator; a number of these "hypthetical" gene structures may in fact be false, again distorting our assessment of the predictive accuracy of a new gene finder when it is tested against these annotated gene structures. In the case of combiner-type programs, a further possibility for bias their evaluation exists—namely, the fact that many gene annotations in curated gene sets derive from annotation pipelines that are effectively combiner programs themselves, so that a combiner program under evaluation is effectively assessed by the degree to which the program agrees with some other combiner-like program upon which the human annotators (if any) heavily depended during genome annotation.

In order to improve this situation, a set of standardized gene sets—more than one, and ideally more than a few—need to be generated and rigorously maintained as new isoforms of existing genes are discovered. Such standard test sets should come from a variety of organisms, and should also be accompanied by corresponding training sets. Large-scale gene-finder competitions (e.g., *GASP* [53], *EGASP* [16]) whichattempt to evaluate and rank sets of gene finders on a common test set generally do not (and, out of practical reasons, typically cannot) control for the difference in training sets used by the authors of the various programs, even though it has been well-documented that the details of the training regime applied to a particular gene finder can significantly affect the accuracy of the resulting system [23]. More generally, the practice of comparing different gene-finding algorithms by applying completely different software systems embedding those approaches to a common test set fails to account for the many minute modeling decisions which are made by different software authors in implementing their highly complex software systems. Thus, a comparison between program X implementing a model of type M_X and a program Y implementing a different class of model M_Y may be so severely influenced by implementation details of the two software systems as to invalidate, or at least distort, any conclusions which are drawn about the fundamental capabilities of methods M_X and M_Y. The ideal scenario for comparing algorithmic and modeling approaches would involve the implementation of the alternative approaches within the same software code-base, so that differences in accuracy between the different versions of a single software system utilizing different gene-finding strategies may be less influenced by implementation details (e.g., [17]); ideally, such single-code-base experiments should be replicated across several independently-developed code-bases. The availability of larger numbers of open-source gene-finding software systems will hopefully make the latter types of experiments more feasible.

5 Summary and Conclusions

We have reviewed the major approaches currently in popular use for automated gene prediction in eukaryotic DNA. While much progress has certainly been made over the past two decades in building accurate gene-parsing systems, much room yet remains for progress. We have enumerated a number of shortcomings inherent in current state-of-the-art systems, and suggested a number of very broad avenues for possible future research. We have focused in particular on the existence of alternative splicing in mammalian genomes, since the existence of potentially many uncharacterized alternative splice forms in human genes poses a potential barrier to biomedical advances aimed at improving human health. To the extent that alternative splicing is still not adequately addressed by current gene-finding systems, the need for creative proposals for the advancement of the field should be manifestly clear.

References

1. Davuluri RV, Grosse I, Zhang MQ (2001) Computational identification of promoters and first exons in the human genome. *Nature Genetics* 29:412-417.
2. Viterbi A (1967) Error bounds for convolutional codes and an assymptotically optimal decoding algorithm. *IEEE Transactions on Information Theory*, 260-269.
3. Dempster A, Laird N, Rubin D (1977) Maximum likelihood from incomplete data via the EM algorithm. *Journal of the Royal Statistical Society (Series B)* 39:1–38.
4. Rabiner LR (1989) A tutorial on hidden Markov models and selected applications in speech recognition. *Proceedings of the IEEE* 77:257-286.
5. Kulp D, Haussler D, Reese M, Eeckman F (1996) A generalized hidden Markov model for the recognition of human genes in DNA. *ISMB '96*.
6. Majoros WM, Pertea M, Delcher AL, Salzberg SL (2005) Efficient decoding algorithms for generalized hidden Markov model gene finders. *BMC Bioinformatics* 6:16.
7. Salzberg SL, Pertea M, Delcher AL, Gardner MJ, Tettelin H (1998) Interpolated Markov models for eukaryotic gene finding. *Genomics* 59:24-31.
8. Staden R (1984) Computer methods to locate signals in nucleic acid sequences. *Nucleic Acids Research* 12:505-519.
9. Zhang MQ, Marr TG (1993) A weight array method for splicing signal analysis. *Computer Applications in the Biosciences* 9:499-509.
10. Altschul SF, Madden TL, Schaffer AA, Zhang J, Anang Z, Miller W, Lipman DJ (1997) Gapped BLAST and PSI-BLAST: a new generation of protein database search programs. *Nucleic Acids Research* 25:3389-3402.
11. Alexandersson M, Cawley S, Pachter L (2003) SLAM: Cross-species gene finding and alignment with a generalized pair hidden Markov model. *Genome Research* 13:496-502.
12. Majoros WM, Pertea M, Salzberg SL (2005) Efficient implementation of a generalized pair hidden Markov model for comparative gene finding. *Bioinformatics* 21:1782-1788.
13. Felsenstein J (1981) Evolutionary trees from DNA sequences. *Journal of Molecular Evolution* 17:368-376.
14. Durbin R, Eddy S, Krogh A, Mitchison G (1998) *Biological sequence analysis*. Cambridge University Press.
15. Siepel A, Haussler D (2004) Computational identification of evolutionarily conserved exons. *RECOMB'04*, March 27-31, 2004, San Diego.
16. Guigó R, Flicek P, Abril JF, Reymond A, Lagarde J, Denoeud F, Antonarakis S, Ashburner M, Bajic VB, Birney E, Castelo R, Eyras E, Gingeras TR, Harrow J, Hubbard T, Lewis S, Ucla C, Reese MG (2006) EGASP: The human ENCODE genome annotation assessment project. *Genome Biology* 7(Suppl 1):S2.
17. Allen JE, Majoros WH, Pertea M, Salzberg SL (2006) JIGSAW, GeneZilla, and GlimmerHMM: puzzling out the features of human genes in the ENCODE regions. *Genome Biology 7(Suppl 1):S9*.
18. Bahl LR, Brown PF, de Souza PV, Mercer RL (1986) Maximum mutual information estimation of hidden Markov model parameters for speech recognition. In: *Proceedings of the International Conference on Acoustics, Speech and Signal Processing* 1986, pp 49-52.
19. Reichl W, Ruske G (1995) Discriminative training for continuous speech recognition. In: Proceedings of the Fourth European Conference on Speech Communication and Technology (EUROSPEECH-95): 18-21 September 1995; Madrid. Amsterdam: Institute of Phonetic Sciences. pp 537-540.

20. Normandin Y (1996) Maximum mutual information estimation of hidden Markov models. In: *Automatic Speech and Speaker Recognition*. Lee C-H, Soong FK, Paliwal KK (eds). Klewer Academic Publishers, Norwell. pp 58-81.
21. Krogh A (1997) Two methods for improving performance of an HMM and their application for gene finding. In: *Proceedings of the Fifth International Conference on Intelligent Systems for Molecular Biology*. Gaasterland T, Karp P, Karplus K, Ouzounis C, Sander C, Valencia A (eds). American Association for Artificial Intelligence. pp 179-186.
22. Gross SS, Brent MR (2005) Using multiple alignments to improve gene prediction. *RECOMB'05*. pp 374-388.
23. Majoros WM, Salzberg SL (2004) An empirical analysis of training protocols for probabilistic gene finders. *BMC Bioinformatics* 5:206.
24. Vinson J, DeCaprio D, Luoma S, Galagan JE (2006) Gene prediction using conditional random fields (abstract). In: *The Biology of Genomes*, Cold Spring Harbor Laboratory, New York, May 10-14, 2006.
25. Culotta A, Kulp D, McCallum A (2005) Gene prediction with conditional random fields. *Technical Report UM-CS-2005-028*. University of Massachusetts, Amherst.
26. Fariselli P, Martelli PL, Casadio R (2005) The posterior-Viterbi: a new decoding algorithm for hidden Markov models. *BMC Bioinformatics* 6 Suppl 4:S12.
27. Käll L, Krogh A, and Sonnhammer ELL (2005) An HMM posterior decoder for sequence feature prediction that includes homology information. *Bioinformatics* 21 Suppl. 1, i251-i257.
28. Stanke M, Waack S (2003) Gene prediction with a hidden Markov model and a new intron submodel. *Bioinformatics* 19:II215-II225.
29. Korf I (2004) Gene finding in novel Genomes. *BMC Bioinformatics* 5:59.
30. Castellano S, Lobanov AV, Chapple C, Novoselov SV, Albrecht M, Hua D, Lescure A, Lengauer T, Krol A, Gladyshev VN, Guigó R (2005) Diversity and functional plasticity of eukaryotic selenoproteins: Identification and characterization of the SelJ family. *Proc Natl Acad Sci* 102:16188–16193.
31. Delcher A, Harmon D, Kasif S, White O, Salzberg SL (1999) Improved microbial gene identification with GLIMMER. *Nucleic Acids Research* 27:4636-4641.
32. Shmatkov AM, Melikyan AA, Chernousko FL, Borodovsky M (1999) Finding prokaryotic genes by the 'frame-by-frame' algorithm: targeting gene starts and overlapping genes. *Bioinformatics* 15:874-886.
33. McCauley S, Hein J (2006) Using hidden Markov models and observed evolution to annotate viral genomes. *Bioinformatics* 22:1308-1316.
34. Misra S, Crosby MA, Mungall CJ, Matthews BB, Campbell KS, Hradecky P, Huang Y, Kaminker JS, Millburn GH, Prochnik SE, Smith CD, Tupy JL, Whitfied EJ, Bayraktaroglu L, Berman BP, Bettencourt BR, Celniker SE, de Grey AD, Drysdale RA, Harris NL, Richter J, Russo S, Schroeder AJ, Shu SQ, Stapleton M, Yamada C, Ashburner M, Gelbart WM, Rubin GM, Lewis SE (2002) Annotation of the Drosophila melanogaster euchromatic genome: a systematic review. Genome Biology 3:RESEARCH0083.
35. Thanaraj TA, Stamm S, Clark F, Riethoven JJM, Le Texier V, Muilu J (2004) ASD: the Alternative Splicing Database. *Nucleic Acids Research* 32:D64-D69.
36. Wojtowicz WM, Flanagan JJ, Millard SS, Zipursky SL, Clemens JC (2004) Alternative splicing of Drosophila Dscam generates axon guidance receptors that exhibit isoform-specific homophilic binding. *Cell* 118:619-33.
37. Parra G, Reymond A, Dabbouseh N, Dermitzakis ET, Castelo R, Thomson TM, Antonarakis SE, Guigo R (2006) Tandem chimerism as a means to increase protein complexity in the human genome. *Genome Research* 16:37-44.

38. Cawley SE, Pachter L (2003) HMM sampling and applications to gene finding and alternative splicing. *ECCB 2003*:36-41.
39. Dror G, Sorek R, Shamir R (2004) Accurate identification of alternatively spliced exons using support vector machines. *Bioinformatics* 21:897-901.
40. Yeo GW, Van Nostrand E, Holste D, Poggio T, Burge CB (2005) Identification and analysis of alternative splicing events conserved in human and mouse. *PNAS* 102:2850-2855.
41. Rätsch G, Sonnenburg S, Schölkopf B (2005) RASE: recognition of alternatively spliced exons in C.elegans. *Bioinformatics* 21 Suppl 1:i369-377.
42. Ohler U, Shomron N, Burge CB (2005) Recognition of unknown conserved alternatively spliced exons. *PLoS Computational Biology* 1:113-22.
43. Wang Z, Rolish ME, Yeo G, Tung V, Mawson M, Burge CB (2004) Systematic identification and analysis of exonic splicing silencers. Cell 119:831-845.
44. Pertea M, Salzberg SL (2002) Computational gene finding in plants. *Plant Molecular Biology* 48:49-48.
45. Uberbacher EC, Mural RJ (1991) Locating protein coding regions in human DNA sequences using a multiple-sensor neural network approach. *PNAS* 88:11261-11265.
46. Vapnik V (1998) *Statistical Learning Theory*. John Wiley and Sons.
47. Jaakkola TS, Haussler D (1999) Exploiting generative models in discriminative classifiers. *Advances in Neural Information Processing Systems* 11:487-493.
48. Zien A, Rätsch G, Mika S, Scholkopf B, Lengauer T, Muller K-R (2000) Engineering support vector machine kernels that recognize translation initiation sites. *Bioinformatics* 16:799-807.
49. Sun YF, Fan XD, Li YD (2003) Identifying splicing sites in eukaryotic RNA: support vector machine approach. *Comput Biol Med.* 33:17-29.
50. Bedell JA, Korf I, Gish W (2000) MaskerAid: a performance enhancement to RepeatMasker. *Bioinformatics* 16:1040-1041.
51. Heber S, Alekseyev M, Sze SH, Tang H, Pevzner PA (2002) Splicing graphs and EST assembly problem. *Bioinformatics* 18 Suppl 1:S181-8.
52. Karolchik D, Baertsch R, Diekhans M, Furey TS, Hinrichs A, Lu YT, Roskin KM, Schwartz M, Sugnet CW, Thomas DJ, Weber RJ, Haussler D, Kent WJ (2003) The UCSC genome browser database. *Nucleic Acids Research* 31:51-54.
53. Reese MG, Eeckman FH, Kulp D, Haussler D (1997) Improved splice site detection in Genie. *Journal of Computational Biology* 4:311-323.

Enhancing Coding Potential Prediction for Short Sequences Using Complementary Sequence Features and Feature Selection

Yvan Saeys and Yves Van de Peer

Department of Plant Systems Biology, Ghent University,
Flanders Interuniversity Institute for Biotechnology (VIB),
Technologiepark 927 B-9052, Ghent, Belgium
{yvan.saeys,yves.vandepeer}@psb.ugent.be

Abstract. The identification of coding potential in DNA sequences is of major importance in bioinformatics, where it is often used to assist expert systems that automatically try to recognize genes in genomes. For longer sequences, the identification of coding potential tends to be easier due to a better signal-to-noise ratio, whereas for very short sequences the issue becomes more problematic. In this paper, we present new methods that specifically aim at a better prediction of coding potential in short sequences. To this end, we combine different, complementary sequence features together with a feature selection strategy. Results comparing the new classifiers to state of the art models show that our new approach significantly outperforms the existing methods when applied to short sequences.

1 Introduction

An important task in current bioinformatics is the analysis of newly sequenced genomes. A first step in this process is the identification of the exact location and structure of the genes in the genome, often referred to as gene prediction. Within expert systems for gene prediction, systems to predict the coding potential for a given subsequence are of high importance. Coding potential prediction is the problem of assessing the probability that a given DNA subsequence encodes a (part of a) protein. As large parts of most genomes do not encode proteins (in the Human genome e.g. only about 5% of the genome codes for proteins), coding potential prediction methods are important to locate the informational parts of the sequence.

Current gene prediction programs are complicated frameworks that combine different submodels together with an optimization technique to find the most likely gene structure. The most important techniques that are used nowadays to model the whole gene prediction framework are based on hidden Markov models (HMM; [11,18]). These models typically combine the results of several submodels, each performing their own, specific task. The submodels of a gene predictor can be divided into two classes: content sensors and signal sensors. While signal sensors are intended to recognize specific functional sites (start and

K. Tuyls et al. (Eds.): KDECB 2006, LNBI 4366, pp. 107–118, 2007.
© Springer-Verlag Berlin Heidelberg 2007

Fig. 1. An unknown DNA sequence can be translated into an amino acid sequence in three possible ways (reading frames)

stop codons, splice sites), content sensors are used to get an impression of the composition of a particular subsequence. This is used to discriminate between protein coding regions and non coding sequences. The ability to discriminate between coding and non coding sequences relates to the structure of coding sequences, which are organized in codons, and these codons are not equally used (codon bias). Whereas for longer sequences it is easier to detect this codon bias, the issue becomes more problematic when short sequence fragments need to be analyzed.

To detect the peculiarities of coding sequences, a large number of protein coding measures were developed [4]. In this paper, we provide a unifying framework that combines many of these measures, and integrates them into a single, discriminative, classification framework.

The rest of the paper is organized as follows. We start by summarizing the existing techniques for coding potential prediction. Subsequently we elaborate on how features can be extracted from these different techniques, and how we integrate them into a new classification model. As many of the included features can be expected to be redundant or irrelevant, we explore the effect of feature selection, comparing the new methods to existing techniques. We end with some concluding remarks and future perspectives.

2   Current Techniques for Coding Potential Prediction

To discriminate between DNA sequences that code for proteins and sequences that do not, we need some characteristics (features) that are typical to coding sequences. As coding sequences are translated into amino acids (the constituents of proteins), and this translation occurs by converting three subsequent DNA nucleotides (termed a codon) into one amino acid, coding sequences will have a specific codon structure. When confronted with an unknown DNA sequence, there are three possible ways in which it can be translated into amino acids (Figure 1). Each of these possibilities is called a reading frame, hence the reading frame determines on which position in the sequence the codons start. Coding sequences are known to have a codon bias, i.e. certain codons (subwords of length three) occur more often than others, and this bias is different from non coding sequences. To capture these differences in composition between coding and non coding sequences, a number of techniques have been developed. These can be roughly divided into two classes: methods based on Markov models, and methods based on signal processing.

2.1 Markov Models for DNA Sequence Modeling

Markov models are generative inductive classifiers, i.e. they learn a model of the joint probability $p(S, c)$ of the sequences S and the label c (where c can be either coding or non coding), and make their predictions by using Bayes' rule to calculate $p(c|S)$, and then picking the most likely label c. Using Bayes' rule, the probabilities $p(c|S)$ can be calculated from the class-conditional probabilities $p(S|c)$. For a kth order Markov model, the probabilities $p(S|c)$ are modeled as:

$$p(S|c) = p(s_1|c) \cdots p(s_k|s_1, ..., s_{k-1}, c) \cdot \prod_{i=k+1}^{n} p(s_i|s_{i-k}, ..., s_{i-1}, c)$$

where s_i denotes the ith character in sequence S, and c is one of {coding, non coding}. This means the Markov model uses a sliding window approach over the sequence, where positions are scored, depending on at most k previous positions. When confronted with a new sequence S', the probabilities $p(\text{coding}|S')$ and $p(\text{non coding}|S')$ are calculated using the Markov model, and the most likely class is chosen as the prediction.

If higher order dependencies between nucleotides exist, then modeling these should result in better prediction methods, as they model the underlying sequence in a more realistic fashion [2]. However, to model DNA, the number of parameters of the model (the k-mer probabilities) increases exponentially with the order k of the Markov model: $O(4^{k+1})$. As a result, the number of training instances needed to reliably estimate each model parameter also grows exponentially with k. A more common way to account for a limited number of training examples is the Interpolated Markov Model (IMM, [16]). This method combines probabilities from contexts of varying lengths (i.e. various orders) to make predictions. It then uses only those contexts for which sufficient data is available. For a given order k, this is done by estimating the probabilities as weighted linear combinations of the parameters $p(s_i|c), p(s_i|s_{i-1}, c), \cdots, p(s_i|s_{i-k}, \cdots, s_{i-1}, c)$. The weights of the linear combination are computed as a combination of two parameters: the χ^2 significance and the frequency of occurrence. As a result, higher order interactions will only be modeled if there is enough training data and if including them provides a significant difference compared to using only the lower order combinations.

2.2 Methods Based on Signal Processing

Methods from signal processing are often used to find regularities in a signal. When applied to genomic sequences, the general notion of "time" in signal processing is interpreted as "position" in the sequence, and regularities can be found by converting the sequence into a numerical format, and subsequently applying signal processing techniques.

For coding potential prediction, the signal processing technique that is most often used is the Fourier transform [17]. Other transforms that can be used are the "run" transform [4] and more recently also the Z-transform [5].

$$
\begin{array}{llllllllllll}
\text{Sequence} & \text{G} & \text{C} & \text{T} & \text{G} & \text{A} & \text{T} & \text{C} & \text{G} & \text{A} & \text{T} \\
\text{Apply } U_A & 0 & 0 & 0 & 0 & 1 & 0 & 0 & 0 & 1 & 0 \\
\text{Apply } U_T & 0 & 0 & 1 & 0 & 0 & 1 & 0 & 0 & 0 & 1 \\
\text{Apply } U_C & 0 & 1 & 0 & 0 & 0 & 0 & 1 & 0 & 0 & 0 \\
\text{Apply } U_G & 1 & 0 & 0 & 1 & 0 & 0 & 0 & 1 & 0 & 0
\end{array}
$$

Fig. 2. Example of converting a DNA sequence to numerical sequences

The most common way to apply Fourier analysis to DNA sequences is to decompose them first into four binary indicator sequences, apply the Fourier transform to each of these sequences, and then sum the Fourier coefficients [19].

Following the notation of Voss [21], a binary indicator sequence is obtained by using a projection operator U_α which selects the elements of the sequence that are equal to the symbol α, namely $U_\alpha(x_j) = 1$ if $x_j = \alpha$ and 0 otherwise. Using the operators U_A, U_T, U_C and U_G then results in four binary sequences. An example is given in Figure 2. For each of the indicator sequences, we can then calculate the magnitudes of the Fourier coefficients $\|F_n^\alpha\|$ for each $\alpha \in \{A, T, C, G\}$, and the sum of these magnitudes represents a global measure of periodicity for the given sequence:

$$
\sum_\alpha \|F_n^\alpha\|
$$

The magnitudes $\|F_n^\alpha\|$ are calculated as $\sqrt{F_n^{r\,2} + F_n^{i\,2}}$, where F_n^r and F_n^i denote the real and imaginary part of F_n respectively. The coefficients F_n are calculated with the standard formula of the discrete Fourier transform

$$
F_n = \sum_{k=0}^{N-1} f_k \cos\left(\frac{i 2\pi n k}{N}\right) - i \sum_{k=0}^{N-1} f_k \sin\left(\frac{i 2\pi n k}{N}\right)
$$

where f_k is the value of the indicator sequence at position k and N denotes the length of the signal.

A well known characteristic when applying Fourier analysis to DNA coding sequences is the observation of a peak at frequency $n/N = 1/3$ in the Fourier spectrum [17,21]. The peak at this frequency is a direct result from the fact that coding sequences consist of codons, combined with the fact that the codons are not equally used. As a result, this peak is a recognition of the boundary between codons, rather than a recognition of an exact repeat of a triplet. The latter would not only lead to a peak at frequency[1] $n/N = 1/3$, but also to peaks at its harmonics which are integer multiples of the frequency $1/3$. As a result, the height of the peak at frequency $1/3$ can be chosen as an indicator of the type of sequence one is confronted with: sequences with higher values for this frequency will be more likely to code for proteins than sequences with a low value.

---

[1] To be notationally correct, we should clarify that the unity in this frequency is one in three nucleotides. However, we will just use the abbreviation $1/3$ during the rest of the text.

3   Combining Complementary Sequence Features for Coding Potential Prediction

In this work, we explore a new approach to combine both strengths of Markov models and methods based on signal processing. To this end, we construct a new set of features, derived from these methods and combine them with a discriminative classifier (a linear support vector machine, [3,20]) to discriminate between coding and non coding sequences. In discriminative learning, the posterior $p(c|S)$ is modelled directly, instead of solving a more general problem as in generative learning, where the intermediate step $p(S|c)$ is modelled [12].

In order to incorporate features that capture the characteristics of Markov models, we construct sets of features that represent the composition of the DNA sequences. These features can be represented as k-mers (subwords of length k over the DNA alphabet {A, T, C, G}), each feature denoting the frequency of the given subword in the sequence. Due to the fact that an unknown sequence can be in one of three reading frames we construct both reading frame dependent as reading frame independent feature sets.

In a similar way, features can be derived from signal processing techniques. For each transform, we extract a number of features that capture the transformed signal in the DNA sequence. Combining both types of features results in the following set of 5992 features:

- Markov-based features:
  - Frame-dependent k-mers. For each of the three possible reading frames k-mer frequencies ($1 \leq k \leq 3$) were calculated, resulting in 252 features.
  - In-frame k-mers. Assuming the sequence is in reading frame 0 (start of the sequence coincides with the start of a codon), in-frame k-mer frequencies ($4 \leq k \leq 6$) were calculated, resulting in a set of 5376 features.
  - Frameless k-mers. For each possible k-mer ($1 \leq k \leq 3$), the global frequencies of occurrence are calculated (i.e. without taking into account the reading frame). This results in 84 features.
- Features based on signal processing methods:
  - Features extracted from the Fourier transform: a) for each of the four indicator sequences, the magnitude of the peak at frequency $1/3$ in the Fourier spectrum (4 features, see Voss (1992) for details), b) the global magnitude at frequency $1/3$, which is the sum of all four magnitudes of the indicator sequences (1 feature), and c) the signal to noise ratio of the peak at frequency $1/3$ (1 feature, see Tiwari et al., 1997 for details).
  - Features extracted from the Z transform: the Z curve parameters are calculated for the frequencies of frame-dependent k-mers ($1 \leq k \leq 3$), using the Z-transform of DNA sequences, as exemplified in Gao and Zhang (2004). This results in a set of 189 features.
  - Features extracted from the "run" transform: for each of the nontrivial subsets of {A, T, C, G} a new sequence is constructed by replacing each base present in the subset with a 1 and replacing each base not in the subset with a 0. Using this transform of the sequence, the number of

runs of 1's of length 1, 2, 3, 4, 5 and greater than 5 are then counted.
This results in a set of 84 features [4].
- Additional feature:
  - A feature was added that denotes if the sequence (assuming it is in
    reading frame 0) contains an in-frame stop codon, or not (ORF feature).
    This feature is used as a post-processing step after applying the LSVM
    classification to filter out false positive predictions.

4   Parallel Feature Selection for Coding Potential Prediction

One could, however, wonder if all these features are equally important or necessary to discriminate between coding and non coding sequences. In particular,
when combining all different types of features to model sequence biases, chances
are high that some features may be redundant or irrelevant. In order to investigate this, feature selection was performed. In feature selection, one seeks a minimal subset of relevant features that achieves maximal classification performance.
Benefits of applying feature selection include better classification performance,
faster classification models (because less features have to be taken into account),
smaller databases (less features are needed to describe the training instances),
and the ability to gain more insight into the process that is being modeled.
A good introduction to feature selection can be found in [9] and [7]. Furthermore, feature selection is becoming more and more widespread in bioinformatics
[14,15,6] where it is often a necessity to include a feature selection method in
the setup, in order to obtain optimal results.

In our work, we adopted a Markov blanket based filter approach, introduced
by Koller and Sahami [10]. This algorithm has a solid mathematical basis, and
has the advantage of being fast and taking into account feature dependencies.
The algorithm eliminates features whose information content is subsumed by
some number of the remaining features. The central idea of the algorithm is a
Markov blanket. Let X denote the full set of features, C the class attribute, and
M some set of features that does not contain X_i. Then M is a Markov blanket
for X_i if X_i is conditionally independent of $(X \cup C) - M - \{X_i\}$ given M. An
approximate algorithm is then suggested that, starting from the full feature set,
iteratively removes the feature with the "best" Markov blanket. Figure 3 shows
the pseudo code of the algorithm.

In a first, preparatory step, the cross-entropy of the class distribution given
pairs of features is calculated for every feature pair. In a next step, for each
feature a possible Markov blanket is defined by selecting the K features X_j
for which the class C and X_i are as most conditionally independent as possible given X_j. The parameter K determines the size of the Markov blanket and
exponentially increases running time as K gets larger (in our experiments we
chose $K = 1$). In a second step, the expected cross-entropy $\delta_G(X_i|M_i)$ is used to

Algorithm KS (Koller-Sahami)

1. Calculate the cross-entropy of the class distribution given pairs of
   features
   $\gamma_{ij} = KL(p(C|X_i = x_i, X_j = x_j), p(C|X_j = x_j))$
   of every pair of features X_i and X_j
2. Instantiate G to X and iterate the following steps until some
   pre-specified number of features have been eliminated:
   - For each feature $X_i \in X$, let M_i be the set of K features X_j in
     $G \backslash \{X_i\}$ for which γ_{ij} is smallest
   - Compute $\delta_G(X_i|M_i)$ for each i
   - Choose the i for which this quantity is minimal, and define
     $G = G \backslash \{X_i\}$.

Fig. 3. Pseudo code for the Markov blanket filter approach of Koller and Sahami

approximate how close M_i is to being a Markov blanket for X_i. This quantity is
defined as:

$$\delta_G(X_i|M_i) = \sum_{x_{M_i}, x_i} p(M_i = x_{M_i}, X_i = x_i) \cdot KL_{f_M, f_i}$$

with

$$KL_{f_M, f_i} = KL(p(C|M = x_M, Xi = xi), p(C|M = x_M))$$

where the Kullback-Leibler divergence KL between two distributions μ and σ
over a set Ω is defined as

$$KL(\mu, \sigma) = \sum_{x \in \Omega} \mu(x) \log \frac{\mu(x)}{\sigma(x)}$$

In a last step, the feature for which M_i most closely resembles a Markov blanket
is eliminated and the process is repeated. In the limit, the elimination of features
can be iterated until the empty set of features is reached.

To cope with our large set of features (5992 features) in a reasonable amount
of time, we designed a parallel version of the algorithm. This parallelization is
based on the fact that most of the time in the algorithm is spent in calculating
the matrices of correlations and expected cross-entropies between features. As
the calculation of the expected cross-entropies requires the calculation of the full
feature correlation matrix, we first need to calculate the matrix of correlations.
This matrix calculation (only the upper diagonal part has to be calculated) can
be sped up in a linear way using n processors in parallel. When all n processors
finish the calculation of their part of the matrix, the different matrix results are
gathered, and the second parallel phase of the algorithm starts. In a similar way
as the correlation matrix is calculated, the matrix of expected cross-entropies is
calculated in parallel. Upon completing the calculation of this last matrix, the
last phase of the algorithm is started, which consists of (sequentially) eliminating
the worst feature in an iterative way.

In essence, the Markov blanket feature selection method returns a ranking of the features, which can be used afterwards to eliminate features. As we have no prior knowledge on the size of a good feature set for the problem of coding potential prediction, various feature subset sizes were evaluated, ranging from 10 features to the full set of 5992 features. The following set of feature subset sizes was evaluated: {10, 20, 50, 100, 150, 200, 250, 500, 750, 1000, 1500, 2000, 2500, 5992}.

5 Results

For our experiments, we used data from the Human genome[2]. The following procedure was used to extract the datasets for coding potential prediction. In a first step, the dataset was cleaned by removing genes with wrong start or stop codons, in-frame stop codons, or genes whose length was not a multiple of three. Next, coding exons were extracted as positive learning examples, and introns and UTR sequences were extracted as negative learning examples. As we were interested in analyzing the behavior of algorithms for short exons, all sequences were divided into length classes (similar to Gao and Zhang, 2004): less than 42 nucleotides (nt), [42-63nt[, [63-87nt[, [87-108nt[, [108-129nt[, [129-162nt[, [162-192nt[and 192 or more nt.

For testing purposes we adopted the following strategy. For each length class in the range [42-192nt[, an equal number of negative examples was added to the positive examples to obtain a balanced dataset. In the case of insufficient non coding sequences for a particular length class, additional sequences were extracted from the length class of 192 or more. This yields, for each length class, a balanced dataset, which was independently split five times in half, obtaining five replications of a two-fold crossvalidation to test on. For each of these ten folds, all the other available data (including the data from all other length classes) was used as additional training data, as is traditionally done in the training of Markov models. Table 1 summarizes the number of sequences and training data used for each length class.

The performance of several algorithms was compared for various (short) sequence lengths. To this end, we compared our method (denoted as SVM) to three existing techniques for coding potential prediction: an 8th order Interpolated Markov Model (IMM-8), the Fourier method described by Tiwari using the signal-to-noise ratio of the peak at frequency 1/3 (denoted as SNR [19]), and the recent Z-curve method from Gao and Zhang, denoted as ZCURVE [5].

As input features for the SVM, all features mentioned earlier are combined, resulting in a set of 5992 features describing each sequence fragment. Features that depend on the length were length-normalized, and before training the SVM all features were scaled between 0 and 1. The C-parameter of the SVM was tuned using a 5-fold cross-validation of the training set. For the ORF-feature, we applied a post-processing step to the algorithm, ensuring that sequences with

---

[2] The data we used is publicly available on the TIGR website http://www.tigr.org/software/traindata.shtml

Table 1. Characteristics of the dataset. For each length class, the total number of exons and the total amount of training data (in kiloBases (kB)) is shown.

Length class	# of exons	Training data
[42-63[	838	10,481 кB
[63-87[	1,648	10,481 кB
[87-108[	1,655	10,481 кB
[108-129[	1,592	10,481 кB
[129-162[	2,026	10,481 кB
[162-192[	1,227	10,481 кB

Table 2. Comparison of the classification performance of the different methods. Classifiers were evaluated using two measures: FP-rate at a TP-rate (sensitivity) of 95% (Se95), and the area under the ROC curve (AUC). For each experiment, the mean and standard deviation are shown.

Measure	Length	IMM-8	SNR	ZCURVE	SVM
Se95	[42-63[	35.36 (1.52)	87.82 (1.79)	25.08 (1.54)	12.97 (2.05)
	[63-87[	23.84 (1.03)	83.69 (1.16)	13.46 (1.06)	7.21 (1.78)
	[87-108[	31.29 (1.22)	75.55 (1.15)	14.19 (1.50)	3.12 (1.21)
	[108-129[	12.10 (0.87)	72.72 (1.24)	5.98 (0.83)	1.35 (0.98)
	[129-162[	6.74 (1.06)	56.84 (1.27)	3.69 (1.01)	0.84 (0.86)
	[162-192[	5.08 (0.99)	49.41 (1.25)	2.29 (0.86)	0.22 (0.44)
AUC	[42-63[	89.95 (0.67)	69.17 (1.20)	93.22 (0.86)	96.78 (1.02)
	[63-87[	93.65 (0.65)	74.86 (0.93)	96.91 (0.43)	98.75 (0.74)
	[87-108[	89.06 (0.74)	80.90 (0.70)	96.45 (0.84)	99.41 (0.54)
	[108-129[	97.25 (0.49)	84.00 (0.49)	98.65 (0.39)	99.69 (0.45)
	[129-162[	98.63 (0.43)	88.20 (0.60)	99.24 (0.41)	99.83 (0.37)
	[162-192[	98.99 (0.41)	90.42 (0.58)	99.50 (0.38)	99.96 (0.17)

in-frame stop codons are always predicted as negatives. This was done by setting the output of the SVM to a very large negative value. For the implementation, we made use of the SVM^{light} package [8].

5.1 Comparison of Classifiers

All results were obtained using five replications of a two-fold crossvalidation, as explained earlier. Furthermore, a statistical test (a combined 5x2 crossvalidation F test, [1]) was used to asses the significance of the differences among the algorithms compared. Table 2 shows the classification performance of the different methods. For each length class and method combination, the result shown is the false positive rate (FPR) at a true positive rate (TPR, sensitivity) of 95% (denoted as Se95) and the area under the ROC curve (AUC, [13]).

From the results, it can be observed that the new model (combining features from signal processing and Markov models) outperforms the existing techniques. Moreover, all differences were found to be statistically significant using the combined 5x2 cv F test. The worst results were obtained using the signal-to-noise ratio of the Fourier transform of the DNA sequence (SNR). This can be explained by the fact that this method only looks at the inherent periodicity in the sequence. As this signal grows stronger with increasing sequence length, it is no surprise that the method deteriorates when short sequence fragments are analyzed. The other method based on signal processing (ZCURVE) performs extremely well, and even outperforms the widely used IMM, while at the same time needing far less parameters than the IMM.

These results justify that a combined approach for coding potential prediction provides a better model to detect coding potential in short (and even longer) sequences.

5.2   Effect of Feature Selection

In order to study the effect of feature selection, we compared the original classifier (SVM without feature selection) to a version using the Markov blanket based feature selection. For each of the previously determined fixed feature subset sizes, we constructed models, and evaluated them using the same cross-validation scheme as was used when comparing the other classifiers. The results of this comparison are shown in Table 3. For each length class, we compare the version without feature selection (denoted as SVM) to the best feature subset found (SVM_{FSS_BEST}). For this subset, both the Se95 and the AUC measure are displayed, as well as the size of the optimal feature subset (the number of features it contains). Furthermore, we included a comparison with a minimal subset of features (SVM_{FSS_MIN}) such that the classification performance did not differ significantly from the one obtained by SVM_{FSS_BEST}.

From these results, it can be observed that feature selection is always beneficial in terms of classification performance. Furthermore, it is clear that the classification performance is achieved using only a small fraction of the features, giving evidence for the fact that many features in the combined feature subset were either irrelevant or redundant. Overall, the global feature set could be approximately reduced to a set containing only 8% of the original features.

Features that are highly ranked include ORF and in-frame stop codon frequencies, features related to AT-composition, nucleotide composition at the first codon position (especially nucleotides G and T) and the signal-to-noise ratio of the peak at frequency 1/3 in the Fourier spectrum.

6   Concluding Remarks and Future Work

In this paper we presented a new framework to combine features from signal processing and Markov models for coding potential prediction. We integrated different characteristics of the sequence at the feature level, and used a discriminative classifier to make a decision, thereby combining all the different features. Our results indicate that combining these different characteristics is a good choice, as it significantly increases the classification performance, compared to existing techniques. In a second step, we analyzed the effect of feature

Table 3. Effect of feature selection for coding potential prediction. For each length class, the results without feature selection (denoted as SVM) are compared to two versions using feature selection. One version that shows the results for the best feature subset (denoted as SVM_{FSS_BEST}), and one version for the minimal feature subset for which the classification performance did not differ significantly from the result obtained using the optimal feature subset. For all algorithms the Se95 measure is shown, complemented with the size of the feature subsets for the variants using feature selection.

LENGTH	SVM	SVM_{FSS_BEST}		SVM_{FSS_MIN}	
	SE95	SE95	SIZE	SE95	SIZE
[42-63[	12.97	11.91	1000	13.54	500
[63-87[	7.21	5.09	750	7.19	250
[87-108[	3.12	2.40	1500	2.82	1000
[108-129[	1.35	1.23	1500	1.29	500
[129-162[	0.84	0.48	1500	0.57	1000
[162-192[	0.22	0.22	ALL	0.28	500

selection for this particular problem. We showed that feature selection proves to be very beneficial, both in terms of classification performance and reduction of the number of features needed by the classifier, especially in the case of shorter sequences.

In future work, we will elaborate more on optimizing the performance of the combined classifier, e.g. by using a more complex kernel for the SVM that would enable us to model feature dependencies. From a more biological point of view, future research will focus on the selected features, and their biological relevance.

References

1. Alpaydin, E.: A Combined 5x2 cv F Test for Comparing Supervised Classification Learning Algorithms. Neural Computation **11(8)** (1999) 1885–1892
2. Borodovsky, M., McIninch, J.: Genemark: parallel gene recognition for both dna strands. Computers and Chemistry **17** (1993) 123–133
3. Boser, B., Guyon, I., Vapnik, V.N.: A training algorithm for optimal margin classifiers. Proceedings of COLT (Haussler,D. ,ed.), ACN Press (1992) 144–152
4. Fickett, J., Tung, C.: Assessment of protein coding measures. Nucleic Acids Research **20** (1992) 6441–6450
5. Gao, F., Zhang, C.: Comparison of various algorithms for recognizing short coding sequences of human genes. Bioinformatics **20(5)** (2004) 673–681
6. Guyon, I., Weston, J., Barnhill, S., Vapnik, V.N.: Gene Selection for Cancer Classification using Support Vector Machines. Machine Learning **46(1-3)** (2002) 389–422
7. Guyon, I., Elisseeff, A.: An Introduction to Variable and Feature Selection. Journal of Machine Learning Research **3** (2003) 1157–1182
8. Joachims, T.: Making large-scale support vector machine learning practical. B. Schölkopf, C. Burges, A. Smola. Advances in Kernel Methods: Support Vector Machines, MIT Press, Cambridge, MA (1998)

9. Kohavi, R., John, G.: Wrappers for feature subset selection. Artificial Intelligence **97(1-2)** (1997) 273–324
10. Koller, D., Sahami, M.: Toward optimal feature selection. Proc. Thirteenth International Conference on Machine Learning (1996) 284-292
11. Majoros, W.H., Pertea, M., Salzberg, S.L.: TigrScan and GlimmerHMM: two open source ab initio eukaryotic gene-finders. Bioinformatics **20(16)** (2004) 2878-2879.
12. Ng, A.Y., Jordan, M.: On discriminative vs. generative classifiers: a com-parison of logistic regression and Nave Bayes. Proc. NIPS 14 (2002).
13. Provost, F., Fawcett, T.: Analysis and visualization of classifier performance: Comparison under imprecise class and cost distributions. Proc. Third International Conference on Knowledge Discovery and Data Mining (1997) 43–48
14. Saeys, Y., Degroeve, S., Aeyels, D., Rouzé, P., Van de Peer, Y.: Selecting relevant features for gene structure prediction. Proc. of the Thirteenth Benelearn conference (2004) 103–109
15. Saeys, Y., Degroeve, S., Van de Peer, Y.: Digging into acceptor splice site prediction: an iterative feature selection approach. Proc. Eighth Conference on Principles and Practice of Knowledge Discovery in Databases (PKDD) (2004) 386–397
16. Salzberg, S., Delcher, A., Kasif, S., White, O.: Microbial gene identification using interpolated markov models. Nucleic Acids Research **26** (1998) 544–548
17. Silverman, B., Linsker, R.: A measure of dna periodicity. J. Theor. Biol. **118** (1986) 295–300
18. Stanke, M., Schöffmann, O., Morgenstern, B., Waack, S.: Gene prediction in eukaryotes with a generalized hidden Markov model that uses hints from external sources. BMC Bioinformatics **7, 62** (2006)
19. Tiwari, S., Ramachandran, S., Bhattacharya, A., Bhattacharya, S., Ramaswamy, R.: Prediction of probable genes by fourier analysis of genomic sequences. Comput. Appl. Biosci. **13** (1997) 263–270
20. Vapnik, V.: The nature of statistical learning theory. Springer-Verlag (1995)
21. Voss, R.: Evolution of long-range fractal correlations and 1/f noise in dna base sequences. Phys. Rev. Lett. **68** (1992) 3805–3808

The Net*Gene*rator Algorithm:
Reconstruction of Gene Regulatory Networks

Susanne Toepfer[1], Reinhard Guthke[2], Dominik Driesch[1], Dirk Woetzel[1],
and Michael Pfaff[1]

[1] BioControl Jena GmbH, Wildenbruchstr. 15, D-07745 Jena, Germany
susanne.toepfer@biocontrol-jena.com
[2] Leibniz Institute for Natural Product Research and Infection Biology -
Hans Knoell Institute, Beutenbergstr. 11a, D-07745 Jena, Germany

Abstract. Mathematical models of gene regulatory networks aim to
capture the causal regulatory relationships by fitting the network models
to monitored time courses of gene expression levels. In this paper, the
Net*Gene*rator algorithm is presented that generates mathematical
models in form of linear or nonlinear differential equation systems. The
problem of finding the most likely interactions between genes is solved by
a structure identification method. This can also be effectively supported
by the incorporation of available expert knowledge. Using favorable parameter identification methods from a system identification point of view
allows to fit accurate and sparsely connected models. By the inclusion
of higher order submodels, the algorithm enables the identification of
gene-gene interactions with significantly time delayed gene regulation.

1 Introduction

Gene regulatory networks control biological functions by regulating the level of
gene expression. Discovering and understanding the complex causal relationships
within gene regulatory networks has become a major issue in systems biology,
computational biology and bioinformatics. Today, large-scale measurement technologies open new opportunities to gain so far unavailable information about
the regulatory mechanisms that underlie specific biological processes as e.g. reactions to different developmental and environmental conditions. For example,
DNA microarray experiments today allow to monitor the output of gene regulatory networks by measuring the gene expression levels of thousands of genes.
Gene expression time courses describe the temporal changes of expression levels
that are caused by the dynamic nature of regulatory interactions. Analyzing
such time series data by reverse engineering techniques allows to provide insight
into the dynamic processes and to generate hypotheses of the causal structure
of specific functional modules of gene regulatory networks.

Using data-driven reverse engineering techniques, structural information is
typically inferred by firstly fitting the parameters of a given mathematical model
to the available time series data and subsequently interpreting the resulting
model structure. In order to infer biologically meaningful models, at least the
following conditions have to be met:

K. Tuyls et al. (Eds.): KDECB 2006, LNBI 4366, pp. 119–130, 2007.

- The mathematical model has to provide an acceptable simplification that leads to an adequate description of the regulatory processes for a certain level of abstraction.
- An appropriate identification algorithm must allow to reverse engineer gene regulatory networks by fitting the model output to the time series observations.
- The time series data have to cover the main regulatory effects of a considered gene regulatory network function with respect to both the gene expression levels and the relevant external input signals.

There exists a large amount of model architectures and corresponding identification schemes for the data-based reconstruction of gene regulatory networks. The different modeling approaches address different aspects of the biological mechanisms. Well-known mathematical models are e.g. directed graphs, Bayesian networks, differential equation systems, stochastic models, Boolean networks and rule-based models [1]. All these models can be interpreted as networks of interacting nodes. Each node has a corresponding node function (e.g. conditional probability distribution, Boolean function, weighted sum) that processes the information coming from other nodes or external inputs. The gene-gene interactions are represented by model parameters that quantify the information processing between the nodes. While in principle the models allow the nodes to interact with each other, it is assumed that in regulatory networks the transcription of many genes is controlled only by a limited number of other gene products. Therefore, in this context it is the general aim of identification to estimate the small subset of *relevant model parameters* out of the set of all possible ones. The relevant parameters are those that are required to generate an adequate fit of the model output to the measured time courses. It is assumed that these relevant parameters coincide with the gene-gene interactions of the underlying gene regulatory network.

In general, the results of data-based modeling critically depend on the quality and quantity of the given data. The data from microarray experiments are corrupted by measurement noise of often unknown characteristics and unfavorable signal-to-noise ratios. Also, because of the high costs, the number of available consecutive time points is still strongly limited. Further, there are hundreds or thousands of genes that are monitored simultaneously. Therefore, without the inclusion of additional mathematical or biological constraints, the relevant model parameters cannot be uniquely estimated from the available data [2].

Methods to cope with this problem using mathematical constraints are first resampling of time courses based on interpolation of time series data [3] and second singular value decomposition based methods [4]. Biological constraints can be taken into account by clustering co-expressed genes and subsequent generation of network models based on clustered time courses. Here, co-expressed genes are assumed to be co-regulated by the same processes [5,6,7]. An additional biologically motivated approach, as used in the NetGenerator algorithm, employs search strategies that are directly based on the assumption of limited connectivity between genes [8,9].

In this paper, the Net *Generator* algorithm is described that infers gene regulatory networks from gene expression time series data. The algorithm uses a structure identification method including a search strategy to identify parsimonious models in form of differential equation systems (Sect. 2.1). Model structure (Sect. 2.2) and parameter identification (Sect. 2.3) are performed using appropriate approaches from system identification theory. Searching for a suitable model structure can be supported by the integration of available expert knowledge (Sect. 2.4). The identification algorithm allows to generate models with an accurate fitting to the observed time courses while the models remain sufficiently simple and interpretable. Section 3 provides an overview of some Net *Generator* applications presented in former publications. The focus of this paper is however a detailed description of the Net *Generator* algorithm.

2   The Net *Generator* Algorithm

2.1   Modeling Approach

The Net *Generator* modeling approach is based on systems of either linear or nonlinear differential equations

$$\dot{x}_i(t) = \sum_{j=1}^{q_s} w_{i,j} x_j(t) + \sum_{l=1}^{p} b_{i,l} u_l(t), \tag{1}$$

$$\dot{x}_i(t) = a_i g\left(\sum_{j=1, j\neq i}^{q_s} w_{i,j} x_j(t) + \sum_{l=1}^{p} b_{i,l} u_l(t) + c_i \right) + w_{i,i} x_i(t). \tag{2}$$

Here, the continuous-valued state variable x_i describes the expression level of gene i. The parameters $w_{i,j}$ are the elements of the gene-gene interaction matrix W that has positive entries for inducers, negative entries for repressors and zero entries if there is no impact of gene j on gene i. The input variable u_l represents the lth environmental factor. The parameters $b_{i,l}$ of the input matrix B quantify how the environmental factor u_l affects the expression level x_i. The change in the expression level x_i at each time point depends on a weighted sum of the influential factors. In the nonlinear differential equations (2), the nonlinear function g represents a nonlinear monotonic sigmoidal activation function. a_i and c_i are additional parameters of this model.

The overall model consists of a set of q_s coupled equations (1) and (2). Such a system models the regulatory interactions between q genes with $q \leq q_s$. While each gene expression time series is typically modeled by a single equation, the Net *Generator* algorithm makes it possible to use more than one differential equation for this purpose. The correlated equations increase the dynamic order of the submodel and allow to identify more complex dynamic behavior. The overall model can be subdivided into q submodels or nodes each consisting of the equations that correspond to a single time series (Fig. 1).

Given the model architecture and suitable, directly measured or preprocessed time series data for the gene expressions and the external inputs, the Net-Generator algorithm generates sparse interaction and input matrices W and B. The determination of the relevant non-zero model parameters is based upon a strategy that separates the model structure identification problem from the model parameter identification problem.

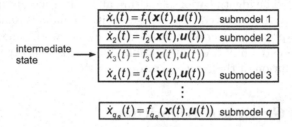

Fig. 1. The model underlying the NetGenerator algorithm: The q submodels that model the q gene expression time series can each consist of one or more differential equations. In case of higher order submodels, intermediate states are included.

2.2 Model Structure Identification

The model structure is given by the links between the network nodes (connectivity) that correspond to the non-zero parameters in the matrices W and B. The NetGenerator structure identification method aims to detect suitable model structures. This is supported by applying favorable parameter identification methods from a system identification point of view.

The NetGenerator algorithm is characterized by the separate identification of the q submodels. In an outer optimization loop, in each iteration step a single time series is identified and the overall model is extended by the newly optimized submodel. However, the given gene expression time series are not identified in a predefined order. Instead this order itself is optimized within an inner optimization loop. Thus, in each iteration step of the inner loop, the identification of all time series (that have not been identified up to this point) is tested and finally the best one selected. Within a single iteration step of the inner optimization loop, the actual submodel structure identification is performed using a search strategy.

The NetGenerator structure identification method therefore consists of three interlocking parts. First, the outer optimization loop carries out the separate identification of the time series. Second, the inner optimization loop optimizes the time series identification order. Third, the submodel structure identification is performed by a search strategy. The following pseudo code illustrates these three parts of the NetGenerator algorithm.

```
NetGenerator algorithm
Input: q time series
Output: Network model consisting of q submodels
% 1st part: outer optimization loop
for actSubmodel=1 to q
    % 2nd part: inner optimization loop
    for actTimeSeries=indexNotYetIdentifiedTimeSeries
        % 3rd part: submodel structure identification
        %            based on a search strategy
        submodels(actTimeSeries)=funIdentification(actTimeSeries)
    end; % inner optimization loop
    bestSubmodel=funSelectBest(submodels)
    improvedBestSubmodel=funImprove(bestSubmodel)
    model(actSubmodel)= improvedBestSubmodel
end; % outer optimization loop
```

Optimization of the time series identification order. The aim of this optimization in the second part is to obtain suitable conditions for parameter identification. This order optimization results in the first selection of simple models that need only few influential factors to be adequately included. In contrast, more complex time series that require many influential factors to be considered in the model are identified later. The advantage of this procedure can be illustrated by the three different types of connections that are present in the interaction matrix W (Fig. 2). Forward connections are positioned in the lower triangular part of the square matrix W. Local feedbacks form the main diagonal of the matrix and backward connections (global feedbacks) form the upper triangular matrix. It should be mentioned that in the final model forward and backward connections have no biological meaning since the order of the final submodels can be arbitrarily permuted. Their order during identification is only relevant with respect to parameter identification.

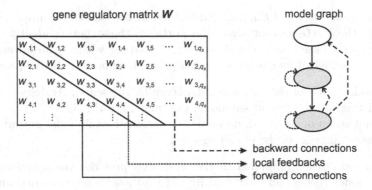

Fig. 2. Forward connections, local feedbacks, backward connections (global feedbacks) with respect to their position in the interaction matrix as well as their direction in the model graph

Now, it is assumed that the submodel for gene i has to be identified and that this gene is regulated by another gene j. Here, the following two cases can be distinguished:

- The time series of gene j has not been identified yet. The corresponding interaction parameter $w_{i,j}$ is an element of the upper triangular matrix, i.e. a backward connection is included. In this case, the time series of gene j is only known at the measured time points. Thus, the simulation of the submodel required for parameter identification (Sect. 2.3) has to rely on interpolated measurement data. Two facts are less favorable here. The measured expression levels are corrupted by noise and the measured time course can considerably differ from the one estimated later. In this case, the estimated parameters and potentially the submodel structure may not be optimal.
- The time series of gene j has already been identified. The parameter $w_{i,j}$ is an element of the lower triangular matrix. That parameter corresponds to a forward connection. For the simulation of the submodel of gene i the simulated time series of gene j instead of the measured one can be used.

Since the measurement data from microarray experiments include considerable noise levels and the model architecture is only a rough simplification of biological regulatory networks, the use of interpolated measurement data for model simulation and parameter identification has disadvantages. Interpolated measurement data are required when backward connections are included into the submodel. Fitting simple time series using only few gene-gene interactions first minimizes the number of critical backward connections during the identification process. Forward connections allow to take into account the actual modeled time courses and their associated modeling errors. The cases described above correspond to the prediction error and the output error approach known from system identification theory [10].

Submodel structure identification. Embedded in the outer and inner optimization loop, the Net*Gene*rator algorithm performs the structure identification of the submodels based on a search strategy. Starting with an initial submodel structure, the algorithm executes the following steps in an iterative procedure:

1. Modification of the submodel structure by the search strategy
2. Fitting of the relevant submodel parameters to the data
3. Simulation of the resulting model to obtain the submodel output
4. Determination of the modeling error

The search strategy applied in the first step provides combinations of influential nodes (genes) and external inputs that are to be examined and compared. All $q + p$ potential influential factors[1] are described by the set $Z = [x_1, \ldots, x_{q_s}, u_1, \ldots, u_p]$.

---

[1] Note that the $q_s - q$ intermediate states that result from submodels with more than one equation are not allowed to influence other nodes.

Testing all possible combinations of influential factors or non-zero parameters is an impractical task even for very small networks. Therefore, the Net*Generator* algorithm employs a heuristic search strategy that makes reasonable restrictions on the search space. Possible solutions are directed towards simple, plausible and interpretable model structures. The search is performed by applying a number of growing and pruning procedures that modify a given submodel structure, e.g. the initial one. The growing and pruning procedures modify the submodel complexity with respect to the number of gene-gene interactions, the number of external inputs and the dynamic order of the submodel. The model selection is controlled by a number of stopping criterions.

Initial submodel. Each submodel structure optimization starts with a simple initial submodel that represents a first order lag element. The initial submodel of gene i has two non-zero parameters: the local feedback parameter $w_{i,i}$ that describes the self-regulation of gene i and the parameter $b_{i,1}$ that quantifies the impact of the first external input on the expression of gene i.

Modification of the submodel complexity. The Net*Generator* algorithm selects subsets of relevant model parameters by searching in two directions: model growing (forward selection) and model pruning (backward elimination). Forward selection is based on the assumption that the best intermediate solution is part of the best final solution. Since this assumption does not have to be true, backward elimination is applied in order to remove unimportant interactions.

1. *Model growing (forward selection):* A forward selection of the most likely interactions is performed by adding new gene-gene interactions or environmental factors. Starting with a given submodel structure with n non-zero parameters, all possible solutions with $n + 1$ non-zero parameters are examined. The best solution with respect to the model fit is retained and further expanded within the next iteration until a stopping criterion is met.

    Selecting gene-gene interactions, forward connections are preferred while backward connections are only included if other connectivities cannot provide acceptable solutions.

2. *Model pruning (backward elimination):* Backward elimination removes gene-gene interactions and environmental factors from the submodel structure. In order to decrease model complexity, all possible solutions that result from the removal of one interaction are considered. Again, the best solution is retained and tested for possible further removals until a stopping criterion is met. If interactions are removed, the algorithm ensures that the decreased model structures remain biologically plausible. For example, structures with only one local feedback parameter are meaningless and are generally excluded.

3. *Inclusion or removal of additional time lag elements:* The third possibility to obtain improved model fits is to adapt the type of dynamic dependency between the interacting genes. The general model structure involves first order dynamics for all submodels. In order to overcome this limitation, the Net*Generator* algorithm allows to include submodels that consist of R differential equations and that represent lag elements of the order R. The search

strategy tests different dynamic orders up to a predefined maximum dynamic order and selects the best fitting one. Although the dynamic behavior of the included higher order submodels can differ significantly, their allowed parameterization is strongly restricted to transfer functions with R equal poles and no or only one zero. Higher order submodels are well suited to identify regulatory interactions that are characterized by significant time delays. They preserve the connectivity of the network model. It should be mentioned that oscillating submodels are not taken into account since the associated submodel complexity would allow them to adapt to highly complex time courses solely based on submodel dynamics instead of submodel connectivity.

Stopping criterion. In order to avoid overfitting and to reach specific user-defined model characteristics, the inclusion or removal of interactions is tied to a number of conditions: (i) An increase in submodel complexity must lead to a considerably improved model fit. (ii) A decreased submodel complexity must lead only to a marginally worsened model fit. (iii) The number of relevant submodel parameters must be smaller than the number of data points in the corresponding time series. (iv) The number of submodel interactions must not exceed a predefined limit.

2.3   Model Parameter Identification

Model parameter identification for a given submodel structure is a repeatedly executed operation of the algorithm. The parameter identification is carried out using a constrained nonlinear optimization based on gradient methods. Here, the mean square error between the model output and the expression data is minimized. All local feedback parameters $w_{i,i}$ are constrained by the condition $w_{i,i} < 0$, i.e. the generated submodels are locally stable. It should be mentioned that even for the linear differential equation systems nonlinear optimization is applied. Linear regression methods require information about the time derivatives. However, estimating the time derivatives from sparsely sampled and noisy time courses is extremely unreliable. This problem coincides with the unfavorable optimization of submodels that include backward connections. Nevertheless, the time derivatives are used for parameter initialization, since linear least squares regression is applied to obtain the initial parameters for the nonlinear optimization. Here, time derivatives are calculated based on a Hermite interpolation. These time derivatives are only used to find initial parameter values. The parameter initialization of the nonlinear model according to (2) is done in the same way. Its additional parameters are initialized so that they provide a linear submodel output for a wide operating range. The initial conditions $x(0)$ are not subject to an optimization method; they are directly derived from the measured time courses.

2.4   Integration of Expert Knowledge

Because of the high complexity of gene regulatory networks and the serious limitations of the measurement data, it is of great advantage to incorporate into the network model as much biological knowledge as possible. The search strategy is based on the assumption that each gene interacts with only a limited number of other genes. However, the structure identification method also allows to introduce specific expert knowledge about the existence or absence of gene-gene interactions or external inputs. Figure 3 illustrates this kind of information and shows the corresponding model graph. The structure identification algorithm ensures that all examined submodel structures are consistent with the expert knowledge put into the model. The possibility to constrain the given model structure can also be used to test different hypotheses extracted in former network reconstruction studies, i.e. the effects of different hypotheses on the model structure can be assessed.

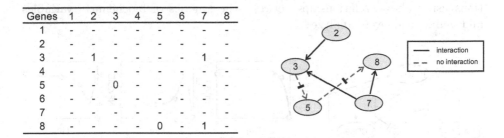

Genes	1	2	3	4	5	6	7	8
1	-	-	-	-	-	-	-	-
2	-	-	-	-	-	-	-	-
3	-	1	-	-	-	-	1	-
4	-	-	-	-	-	-	-	-
5	-	-	0	-	-	-	-	-
6	-	-	-	-	-	-	-	-
7	-	-	-	-	-	-	-	-
8	-	-	-	-	0	-	1	-

Fig. 3. Prior knowledge about interactions between eight genes with the corresponding model graph (- no prior knowledge, 1 interaction exists, 0 interaction does not exist)

3 Applications

The Net*Gene*rator algorithm has been applied to generate hypotheses about gene regulatory interactions of different biological networks from gene expression time series data. In all cases, an comprehensive clustering analysis was performed. The main kinetics were extracted by primarily fuzzy clustering the differentially expressed genes. Clustering analysis included the optimization of the number of clusters by evaluating cluster validity indices and using specific knowledge from databases. Network reconstruction was performed by detecting the regulatory interactions between cluster-representative genes. The selection of these representative genes was based on their fuzzy membership degree to clusters, on biological expert knowledge and also on methods such as gene description text mining. As a result of clustering, the Net*Gene*rator algorithm discovered the relationships between 4 and 10 main kinetics characterizing the biological processes. The available time series contained between 5 and 9 measurements. For several applications, alternative network models were generated and analyzed based on different initializations of the algorithm and the integration of different prior knowledge. For some applications, the robustness of network reconstruction was analyzed by bootstrap studies performing a large number of identification runs with artificially perturbed data. The resulting models included between 8 and 22 interaction parameters that were compared with knowledge not yet included into the network. The following applications have been published:

- Immune response of peripheral blood mononuclear cells to bacterial infection with heat-killed pathogenic *E. coli* [2] (Fig. 4)
- Stress response during recombinant protein expression in *E. coli* [11]
- Effect of LiCl stimulation on hepatocytes [12]
- Stress response to a temperature shift in *A. fumigatus* [13]
- Effect of culture media on primary mouse hepatocytes [14]

In [15], the use of the Net*Gene*rator algorithm for network model based analysis of a bioartificial liver cell system is reported. In contrast to the applications listed above, relationships between the kinetics of biochemical variables and amino acids were analyzed.

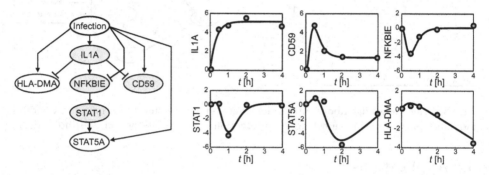

Fig. 4. Graph and simulated output of the inferred network model in [2]

4   Discussion

The Net*Gene*rator algorithm presented in this paper has been devised to reverse engineer gene regulatory network models consisting of differential equations. The advantages of these models lie in their capability to explicitly present dynamic system behavior and to model the dynamics at continuous-valued expression levels, rather than just using the two levels on and off. Well-known other identification methods to infer such dynamic systems are least squares methods [3], singular value decomposition based methods [4], genetic algorithms [7], simulated annealing [5] and search strategies [8,16]. Since most of these approaches rely on linear regression methods, they suffer from the drawback of requiring the time derivatives that have to be calculated from sparsely sampled and noisy measurement data. The Net*Gene*rator algorithm proposed employs nonlinear optimization to estimate the parameters of examined submodel structures. Besides the accurate fitting that can be obtained due to favorably set system theoretical conditions, it also allows the optimization of higher order submodels. These submodels enable the identification of significantly time delayed gene regulation. This feature is very important, since it is known that there can be a considerable time delay between the expression of one gene and the observation of its effects [17]. With the inclusion of intermediate states, it is possible to consider a wider range of biologically meaningful dynamic dependencies between genes.

Due to the limitations of available data, it is extremely important to find an adequate subset of genes, gene clusters or cluster-representative genes for network reconstruction. This selection has to include as much expert knowledge as possible. The applications presented in Sect. 3 use clustering in order to pre-process the given time courses. If the quality and quantity of the data is seriously limited and if no further knowledge is available, for two very similar time courses, it does not make sense to derive different connectivities for the corresponding genes. Even though gene-specific information can be lost and co-expressed genes do not have to be regulated by the same biological process, similar expression patterns should still be clustered. The major benefit of clustering is the inherent reduction of dimensionality and noise.

In reverse engineering, in general, a small modeling error alone provides no guarantee that the model obtained will show structural equivalence to the actual gene regulatory network analyzed. Differential equation systems are rough representations of biological processes, since the complex regulatory effects of intermediate products are simplified to linear or specific nonlinear dynamic relationships between genes [18]. Further, data delivered by DNA microarray experiments in most cases do not contain enough information to reconstruct more complex models. Therefore, as long as data availability is not considerably improved, the resulting models will not be adequate to make precise predictions with respect to the gene regulatory response to modified experimental conditions.

For these reasons, the Net*Generator* algorithm allows to generate hypotheses on the most likely activating or repressing interactions of the underlying network. In order to provide further support for these hypotheses, additional specifically designed experiments are required. Integrating diverse biological knowledge can substantially improve the results of network reconstruction. A limitation of the presented algorithm is its relatively long calculation time for larger networks (e.g. with more than 15 genes) caused by the exhaustive search strategy used within the outer and inner optimization loop. Instead of this search strategy, in general all methods that allow model structure optimization (e.g. genetic algorithms and genetic programming) could be incorporated and combined with the constrained nonlinear optimization. The problem of getting stuck into a local minimum is reduced by using bootstrap techniques and repeated runs of the Net*Generator* algorithm with different configurations. However, the Net*Generator* algorithm presented allows to reconstruct very sparsely connected network models with a high accuracy of model fitting.

References

1. De Jong, H.: Modeling and Simulation of Genetic Regulatory Systems: A Literature Review. Report (2000)
2. Guthke, R., Moeller, U., Hoffmann, M., Thies, F., Toepfer, S.: Dynamic Network Reconstruction from Gene Expression Data applied to Immune Resonse during Bacterial Infection. Bioinformatics 21 (2005) 1626–34
3. D'haeseleer, P., Wen, X., Fuhrmann, S., Somogyi, R.: Linear Modeling of mRNA Expression Levels During CNS Development and Injury. Pacific Symposium on Biocomputing 4 (1999) 41–52
4. Yeung, M. K. S., Tegne, J., Collins, J. J.: Reverse Engineering Gene Networks using Singular Value Decomposition and Robust Regression. Proc. Natl. Acad. Sci. 99 (2002) 6163–68
5. Mjolsness, E., Mann, T., Castano, R., Wold, B.: From Coexpression to Coregulation: An Approach to Inferring Transcriptional Regulation among Gene Classes from Large-Scale Expression Data. Advances in Neural Information Processing Systems 12 (2000) 928–34
6. D'haeseleer, P., Liang, S.,Somogyi, R.: Genetic Network Inference: From Co-Expression Clustering to Reverse Engineering. Bioinformatics 16 (2000) 707–26
7. Wahde, M., Hertz, J.: Coarse-Grained Reverse Engineering of Genetic Regulatory Networks. Biosystems 55 (2000) 129–36
8. van Someren, E.P., Wessels, L.F.A., Reinders, M.J.T., Backer, E.: Searching for Limited Connectivity in Genetic Network Models. International Conference on Systems Biology (2001)
9. Bonneau, R., Reiss, D.J., Shannon, P., Facciotti, M., Hood, L., Baliga, N.S., Thorsson, V.: The Inferelator: An Algorithm for Learning Parsimonious Regulatory Networks from Systems-Biology Data Sets de novo. Genome Biology 7 (2006)
10. Nelles, O.: Nonlinear System Identification. Springer (2001)
11. Schmidt-Heck, W., Guthke, R., Toepfer, S., Reischer, H., Duerrschmid, K., Bayer, K.: Reverse Engineering of the Stress Response during Expression of a Recombinant Protein. Proc. EUNITE Symp. (2004) 407–12
12. Zellmer, S., Schmidt-Heck, W., Gaunitz, F., Baldysiak-Figiel, A., Guthke, R., Gebhardt, R.: Dynamic Network Reconstruction from Gene Expression Data Describing the Effect of LiCl Stimulation on Hepatocytes. Journal of Integrative Bioinformatics (2005)
13. Guthke, R., Kniemeyer, O., Albrecht, D., Brakhage, A.A., Moeller, U.: Discovery of Gene Regulatory Networks in Aspergillus fumigatus. Lecture Notes in Bioinformatics (2006) (submitted)
14. Schmidt-Heck, W., Zellmer, S., Gebhardt, R., Guthke, R.: Effect of Culture Media on Primary Mouse Hepatocytes Identified by Dynamic Network Reconstruction from Gene Expression Data. Lecture Notes in Bioinformatics (2006) (submitted)
15. Schmidt-Heck, W., Zellmer, S., Gaunitz, F., Gebhardt, R., Guthke, R.: Data-based Extraction of Hypotheses about Gene Regulatory Networks in Liver Cells. European Symposium on Nature-inspired Smart Information Systems (2005)
16. Chen, T., He, H.L., Church, G.M.: Modeling Gene Expression with Differential Equations. Pacific Symposium on Biocomputing 4 (1999) 29–40
17. Li, X.,Rao, S., Jiang, W., Li, C., Xiao, Y., Guo, Z., Zhang, Q., Wang, L., Du, L., Li, J., Li, L., Zhang, T., Wang, Q.K.: Discovery of Time-Delayed Gene Regulatory Networks based on Temporal Gene Expression Profiling. Bioinformatics 7 (2006)
18. Wessels, L., van Someren, E., Reinders, M.: A Comparison of Genetic Network Models. Proc. of Pac. Symp. on Biocomputing (2001)

On the Neuronal Morphology-Function Relationship: A Synthetic Approach

Ben Torben-Nielsen, Karl Tuyls, and Eric O. Postma

MICC, Universiteit Maastricht, The Netherlands
B.Torben-Nielsen@micc.unimaas.nl

Abstract. Recent investigations emphasized the role of dendrites in the information processing and computational capabilities of a single neuron. On a local electro physiological level, it is known which computations can be done in dendrites. However, it is still largely unknown how the complete dendritic morphology contributes to the function of a single neuron. In this study we present a synthetic approach to investigate the relationship between morphology and function. Our approach is implemented in a software tool and an experiment is presented. In the experiment we generate morphologies that approximate the functional properties of the *Nucleus Laminaris*. We discuss the possibilities and limitations of our synthesized approach.

1 Introduction

Since the pioneering work of Rall in the 60s, the role of dendrites in neural computation has become more and more prominent (e.g., [12,17,22]). It is observed that the dendritic morphology strongly influences these capabilities, suggesting a relationship between neuronal morphology and the functional capabilities in single neurons.

On a limited local and biophysical level there are several functions and computations known that result from the morphology of neurons. Here, it is important to make the distinction between functions that arise from the passive (intrinsic) properties of neuronal morphology (e.g., lengths and diameters) or functions arising from the active properties (e.g., voltage-gated ion channels) of neurons. We discuss only the function resulting from passive properties, as a passive model of a neuron is a good approximation of the active model [25,17], while keeping the model straight-forward. It is known that passive dendrites can act as delay lines when signals attenuate as they propagate through a dendrite [17]; they can facilitate shunting inhibition by the relative location of inhibitory synapses with respect to the excitatory synapses [23,17]; they can perform local "operations" like logical operations as branching points in the dendritic tree act like a logical ADD function [12]. For a small number of neurons it is known that their morphology mainly optimizes specific connectivity [24]. Neurons whose function is connectivity are beyond the scope of this paper. For the majority of neurons, it is largely unknown how the morphology influences function at the level of a single neuron [22]; at least, in terms of single neuron dendritic morphology influencing

K. Tuyls et al. (Eds.): KDECB 2006, LNBI 4366, pp. 131–144, 2007.

spiking behaviour. Several studies reported strong evidence of a relationship between morphology and function of a single neuron: both in experimental as in computational studies (e.g., [1,25]) is was revealed that different morphologies trigger different neuronal responses (including different neurons within a specific class or type). Thus, the morphology-function relationship is the synthesis of (i) neuronal function, (ii) neuronal morphology, and (iii) electro physiological properties. In this paper we present a synthetic approach to investigate the so-called *morphology-function relationship*.

An experiment is presented in which we try to find a neuronal morphology that exhibits electro physiological properties similar to the *Nucleus Laminaris* neurons. This type of neuron which is located in the auditory brainstem of mainly birds has the function of an auditory coincidence detector, and is one of the few neurons of which the function, morphology, and electro physiological properties are understood [22]. Therefore, this particular type of neuron was suited for our study. The advantage of a synthetic approach over a conventional experimental approach in this type of study is that the investigator has full control over all parameters [11], and is not limited to small availability of biological data [4].

The remainder of this report is outlined as follows. In the next Section (2) we present a motivational, real-life example of the morphology-function relationship. Section 3 introduces our synthesized approach and the experimental set-up of the experiment. Section 4 presents the results while we conclude with discussion of the obtained results and this type of study in Section 5.

2   Morphology-Function Relationship: Motivational Example

The morphology-function relationship is defined as the synthesis of (i) neuronal function, (ii) neuronal morphology, and (iii) electro physiological properties. To clarify this notion, we present a real-life example of the relationship between form and function: neuronal remodelling in holometabolous animals.

Holometabolous animals are animals that go through different life phases which require different behaviours. For instance, the moth (a holometabolous animal) lives as caterpillar (i.e., larval), pupal and adult. The behaviours (i.e., neuronal function) required during these stages are completely different: the slow crawling movement of the caterpillar in contrast to the near oscillatory movement of the wings during flight. The emergence of the appropriate behaviour in the specific life phase is accommodated by morphological changes in the post embryonic brain, the so-called remodelling. Remodelling is the consequence of three structural alterations: neuron death, neuron growth, and reshaping of neurons. The third alteration takes place in the motor neurons which go from slowly firing neurons to oscillatory spiking neurons [8]. Indeed, an interplay between morphological changes and physiological changes was observed. Nevertheless, "there is little known about how the physiological changes accompany structural remodelling" [8]. Yet, as pointed out in [7]: "dendritic remodelling might also be important for modifications of the intrinsic properties of motor neuron", implying that morphological changes affect the passive information processing due to changing intrinsic properties of a dendrite (e.g., [18,29,25]). The last statement enforces the notion of the morphology-function relationship as it is empirically proven that a neuron exhibits a certain behaviour with a certain shape, exhibits different behaviour when morphologically changed.

3   Synthetic Approach and Methods

We adopt the synthetic approach for two reasons. First, in an synthetic experimental set-up the experimenter has full control over all the parameters. Consequently, it is a straight-forward process to make little changes and investigate the influence of these small alterations. With biological data and living tissues the level of control is too little to explore all possibilities on short time-scale. Second, if biological data is available there is no vast amount as to perform "proper" statistics. This issue is especially problematic when concerning morphometric data [2]. In our synthetic approach we use *virtual neurons*, i.e., digitized representation of neurons with an emphasis on their morphology.

In our synthetic approach we combine descriptive and computational modelling. Descriptive modelling does not include any underlying mechanisms and is used here for the generation of virtual neurons. Several tools show the success of descriptive modelling to obtain realistic virtual morphologies, e.g., L-Neuron [5], Neuron PRM [15] , and NeuGen [9]. Computational modelling is more detailed and incorporates underlying mechanisms, albeit abstracted for the sake of tractability. In our approach computational modelling is used to perform electro physiological simulations. The remainder of this section first presents a tool - EvOL-Neuron - we developed that implements our approach. Then, we present an experiment in which we explore a mapping between function and morphology.

3.1   EvOL-Neuron

Our synthetic approach is based on the principle that we can generate *all* neuronal structures (i.e., 3D tree structure with bounded size) and explore these structures as to find structures that obey predefined criteria. This principle is implemented in a tool called EvOL-Neuron [27]. EvOL-Neuron consists of two (intertwined) phases: generation and optimization. In the first phase, candidate virtual neurons are generated. The second phase then searches for specific structures. The two phases are explained below.

First, we use L-Systems to generate virtual neuron morphologies. L-System is a mathematical formalism of rule rewriting named after its inventor Aristid Lindenmayer [16,20]. Originally designed for describing the branched structures of plant morphology, it is highly suitable for describing virtual neuron morphology. The idea of L-Systems is simple and powerful: an axiom defines the initial starting point, and the production rules define how to rewrite the axiom. In cycles, all the symbols currently stored in the L-System are rewritten according to the production rules. This rewriting process is repeated a predefined number of times and results in a long, iteratively built string. L-Systems in itself is nothing more than a way of generating a long string from a parsimonious description. In essence, L-Systems have no semantics. A meaning is given by an interpretation scheme. In the case of virtual neurons, an L-system needs to be interpreted as a 3D structure (resembling neuron morphology). We adopt the Rotation-Elevation interpretation to create 3D structures. In this interpretation, three-dimensional directions are represented in terms of two angles: the rotation and elevation angles [27]. Figure 1 illustrates how a complex structure is generated from a small L-System description. The description used to generate the structures is displayed in the inlay and contains one axiom (axiom_0) and one rule (A). The resulting structure is illustrated after one to four rewriting cycles and illustrates the increasing complexity with more rewriting cycles. In out study we always use four rewriting cycles. This number is chosen as less cycles result in structures with insufficient complexity while more cycles are computationally too expensive to use in the optimization phase.

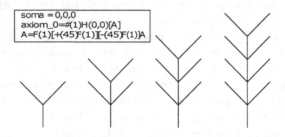

Fig. 1. Illustration of the geometric interpretation. The figures illustrate the development of a structure after rewriting cycle 1 to 4. The inlay contains the L-System description of the structures. In the description, the soma defines the spatial position of he soma (which is not used yet); axiom_0 is the first (and only) axiom; A is the first (and only) production rule.

Second, the optimization (search and selection) is done with Evolutionary Computation (EC). EC is a pragmatic programming method inspired by Darwinian evolution to explore large solution spaces [10,13,19]. EC exploits the principle of *survival of the fittest*. Applied to virtual neurons this means that generated neurons are the *individuals* in a *population* of candidate virtual neurons. The L-System describing a specific individual is called the *genome* of that individual. In EC, the evolution starts with initially random genomes and hence random morphologies. Now, all evolved individuals are tested with respect to their biological accuracy and are assigned a *fitness*. Similarly with Darwinian evolution, the best individuals are allowed to reproduce (by means of selection, cross-over, and mutation). The application of this artificial evolution ensures that individuals gradually reach the predefined criteria. The evolution is allowed to proceed until a predefined fitness is reached, yielding a neuron that obeys this fitness.

A schematic overview of the methodology underlying EvOL-NEURON is illustrated in Figure 2. In summary, electro physiological properties of real neurons are used as prototype, i.e., the goal of the optimization phase. Virtual neurons are encoded as L-System, compared to the prototype functionality and assigned a fitness. On the basis of the fitness value a new population of virtual neurons is constructed until they match with the biological neuron corresponding to predefined properties[1].

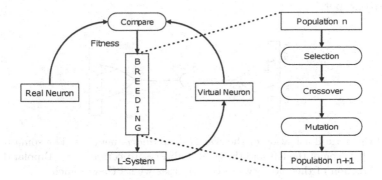

Fig. 2. Evolutionary Computation applied in EvOL-NEURON. In generations (cycles), an initially random set of virtual neurons is refined and compared to predefined criteria. This process terminates when a virtual neuron is found that obeys the predefined criteria or a predefined number of cycles is reached.

3.2 Experiment Description

In the experiment we conducted we tried to automatically find a relationship between function and morphology of the *Nucleus Laminaris* (NL) neurons. Neurons from the NL, which are found in the brain stem of most avian animals, are one of the few neurons of which we understand the morphology-function mapping [22]. NL neurons are bipolar neurons with their two dendrites going to either side - left or right - to connect to the auditory afferents in the projection area (schematically illustrated in Figure 3). When the dendrites make it to the projection area (i.e., area in the brain where the auditory afferents terminate; in a synthetic context referred to as "target zones") they are branching heavily. They fire only when synaptic input is converging at the soma at exactly the same moment; when input is coming from only one of the sides, independent of the strength of this signal, the neuron will not fire.

Thus, these neurons have two phenomenological functional properties. First, they have non-linear summation of synaptic input in the distant part of the dendrites. More specific, the synaptic inputs on a single dendritic branch (i.e., spatially close to each other) will reduce the driving-force in that dendritic branch, and thus saturate. To avoid this reduction in driving force (in order to generate

---

[1] The prototype software can be found at:
   http://www.cs.unimaas.nl/b.torben-nielsen/evol-neuron/main.php

a spike in the soma), synapses need to be spatially distributed over different dendritic branches in the target zone. Despite the fact that too much reduction in driving force must be avoided, it is of high importance that multiple synaptic inputs are summed non-linearly (on a single dendritic branch) as the neuron will otherwise spike from multiple inputs originating from one side. Therefore, we refer to the first property as saturation in single dendritic branches. Second, linear summation of input signals takes place at the soma when input signals are coming from both sides[2]. In this paper we try to find 3D structures that exhibit these two properties.

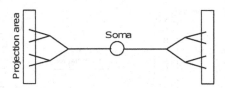

Fig. 3. Schematic illustration of the *Nucleus Laminaris* neurons. The soma (circle) is located in between of the left and the right auditory projection areas. Bipolar dendrites grow in direction of the auditory projection areas where they branch.

3.3 Fitness Assessment

As mentioned earlier in this section, EvOL-Neuron searches for morphologies on the base of survival of the fittest. Here, we explain how we assess the fitness of a specific morphology. As we are interested in finding *any* structure that serves a specific functional (i.e., electro-physiological) goal, we need to run a detailed physiological simulation and use recordings from this simulation to assess a fitness value. As neuronal simulator we used NEURON, a software package that allows the experimenter to set-up a detailed neuronal morphology and run electro-physiological recordings on this morphology [6]. A connection between EvOL-Neuron and NEURON was made by generation in EvOL-Neuron of a script that is readable in NEURON (the so-called *hoc-file*), execute this script by NEURON and capture the output generated by NEURON to process the outputted information further in EvOL-Neuron. The script contains a cell description with morphological details and electro physiological properties of the modelled cell, and, general simulation settings for NEURON. We used a passive model for the dendrites and Hodgkin-Huxley channels for the soma. The number of compartments in each dendritic segment was set to 5, and when a segment intersected with the predefined target zone (see Figure 3) we added 2 synaptic connections to the segment. Synapses were made of Exp2Syn objects in Neuron that were connected to a pulse generator (NetStim object) through the NetCon object. The input consisted of a single pulse generated after 1 ms. The remaining simulation settings we use in NEURON are listed in Table 1.

---

[2] In this paper we present a phenomenological explanation of the NL functions. For a technical description of NL functions we refer the reader to [1,26].

Table 1. Configuration of the simulation in NEURON

E_r	$-65\ mV$
g_pas	0.0001 pS
e_pas	$-65\ mV$
Ra	35.4 Ωcm
Cm	1 pF μm^2
τ rise/decay	0.1 ms, 1 ms
Simulation time	20 ms
dt	0.025 ms

As we want to achieve a specific functionality (i.e., electro-physiological response) we record the membrane potential at the soma. The membrane potential should reflect the two physiological properties of NL neurons: saturation in the dendrites and linear summation in the soma. Saturation in the dendrites is a property directly related to the morphology: synaptic inputs should be spatially distributed or saturation occurs. Therefore, we only need to base the fitness function on the linear summation in the soma. The following formula is used to calculate the fitness value for a given 3D structure (from Stiefel and Sejnowski, personal communication and [26]).

$$F = -\frac{M_r + M_l - M_{rl}}{M_{rl}} - \alpha \left(\frac{M_l}{M_r} + \frac{M_r}{M_l} \right) \beta \log \left(M_{rl} \right)$$

In the above formula M_r, M_l, and M_{rl} are the responses measured in the soma when input was provided on the left side (M_l), right side (M_r) or both sides simultaneous (M_{rl}). This response is measured in three different simulations. Furthermore, α and β are scalars that define the relative importance of the entities in the formula. The fitness function rewards linear summation and balanced input from both sides. There are *no* direct morphological constraints imposed upon the 3D structures. However, generated structures that do not obey *intrinsic* neuronal properties (i.e., bifurcations instead of n-furcations, specific asymmetry and non-increasing diameters away from the soma) are assigned a low fitness and no electro physiological simulation is performed in NEURON. Once a structure is used in a functional simulation we test whether the outcome is realistic: by allowing every 3D tree structure the outcome of a simulation can be unrealistic (e.g., when non-natural changes in diameter occur, the electro physiological response can be highly unpredictable). Unrealistic simulation outcomes also receive a low fitness. The structures that fulfilled these criteria then receive a small reward inverse proportional with their size which introduces a biases towards smaller (i.e., short total length) structures. These extra checks and constraints are required as every 3D structure is considered as a candidate.

4   Results

We performed eight evolutionary runs and they all yielded a good solution according to our fitness function. Here, we discuss the result of one run which is representative for the results of all runs. Figure 4 illustrates the development of the fitness function of the best individual. The fitness values lower than -98 represent a fitness as determined by the heuristics; higher values are computed by the fitness function as given in the previous section.

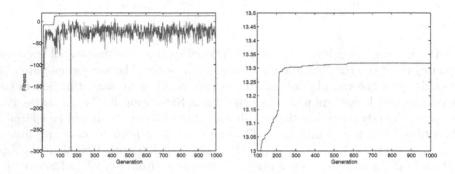

Fig. 4. Fitness development of the best individual of a particular run. Left: top line illustrates the fitness value of the best individual of a generation while the bottom line illustrates the average fitness in a generation. The drop in average fitness (around generation 185) is caused by an erroneous calculation and assessed a extremely low value. Right: detail of the fitness development. The fitness is slowly increasing but no spectacular increase is observed.

We optimized structures to obtain two particular physiological properties: saturation and linear summation at the soma. Figure 5 (left) illustrates the linear summation effect as observed in the generated structure. It is not a perfect summation, but reaches to 88.4% of the sum of both inputs alone. Nevertheless, it is clear that a strong summation is obtained. It must be noted that this result comes from a structure without tapering, as the symbol to update the diameter of a structure was taken out of the evolved description by the genetic algorithm[3]. By means of the addition of two extra symbols to the evolved L-System description we added taper to the resulting structure. Figure 5 (right) illustrates the near-optimal linear summation effect (98.9%) for this structure. In the remainder of this section, we will discuss the result found by the algortihm and not the manually modified result.

The saturation effect can be observed when adding more synapses to the dendritic segments that already had a connection (because they entered the target area). The saturation effect is illustrated in Figure 6. We added 3 and 8 synapses to have a total of 5 and 10 synapses, respectively. As illustrated

---

[3] In our implementation, it is hard to get a specific symbol back in the description once it is taken out. This is due to the particular design of the mutation operator.

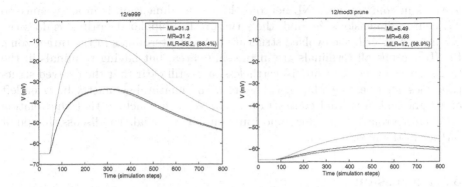

Fig. 5. Linear summation effect measured in the soma. Left: result found in an evolved structure, summation is 88.4% of the sum of the inputs. Right: result found in a manually modified structure; modification was the addition of two symbols (see text), summation is 98.9% of the sum of the inputs.

in Figure 6 (left) this has only a minor effect on the membrane potential in the soma. Only a slightly slower decay is observed when a total of 10 synapses are connected. Figure 6 (right) illustrates the saturation effect in the dendritic segment with the synaptic inputs. Again, a slightly slower decay can be observed for more inputs. Thus, we achieved strong linear summation and nearly perfect saturation.

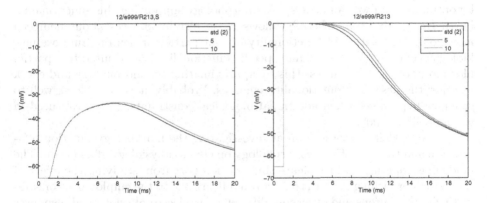

Fig. 6. Saturation effect obtained by a resulting structure. Left: saturation recorded in the soma. Right: saturation recorded in the dendrites. Details in the text.

We obtained a morphology that, when used in electro physiological simulations, obeys our functional requirements. Figure 7 (left) illustrates the evolved neuron that possesses our two predefined electro physiological properties. Figure 7 (right) illustrates the XY projection and the target zones (depicted as lines). It can be observed that the evolved structure does not resemble NL morphology. However, at an abstract level the neuron has three morphological features in common with NL neurons. First, the dimension is in accordance to NL neuron dimensions. Second, the structure has bipolar dendrites as does the NL neuron. Third, the evolved structure has some terminals in the target zone. In NL neurons all terminals are in the same area, but having terminals in the target zone is a prerequisite. Nevertheless, it is still clear that the evolved structure does not resemble a biological neuron: an unnatural symmetry between left and right and abnormal contraction ratio. We can conclude that the relation between neuronal morphology and function is not trivial. We discuss this point further in the next section.

5   Discussion

This paper aims at presenting a new, synthesized approach to study the morphology-function relationship. An import question then arises: *What can we learn from this type of study?* Firstly, supported by our result is that this synthesized approach will provide new insights into morphology. One of these insights is that morphology is not trivial: there is no such thing as one morphology that can produce a specific functional property. Thus, there is no one-to-one mapping in the morphology-function relationship. Moreover, with this synthetic approach (implemented in EVOL-NEURON) we can investigate the targeted relationship in both directions. Morphologies can be generated as to resemble real neurons [27], and the emerging physiological functionality can be analysed. The inverse is possible as well as we demonstrated in this report: functionality can be optimized and we can analyse the morphology supporting this functionality. Furthermore, this type of study allows us to test hypotheses about neuronal morphology in relation to functionality of that particular neuron. Suppose new biological evidence is released and specific functionality is attributed to a particular morphology. We can test this by approximating the morphology and check whether the described functionality emerges. Probably, it is possible as well to generate hypotheses when additional biological constraints can be captured in our modelling study.

Several modelling studies aim at investigating the morphology-function relationship and thus the effect of morphology on electro physiology of a neuron. The modelling approach to investigate this relation goes from studying the effect of theoretical branching patterns on neuronal firing [29] to the employment of complete virtual neurons and observing differences in electro physiological responses after slightly changing the morphology [14]. One study that has the same approach as outlined is this paper (i.e., optimization of morphologies that obey electro physiological requirements) is reported in [26]. They successfully generated virtual neurons for the linear summation task (as used in this paper) and the spike order detection task. In their study there is a striking match between the evolved morphology for the linear summation task and the morphology of Nucleus Laminaris neurons; a match that we did not obtain. The main difference between our study and the study reported in [26] is the algorithm by which a candidate morphology is generated. They use an algorithm based on L-Neuron [3] which uses

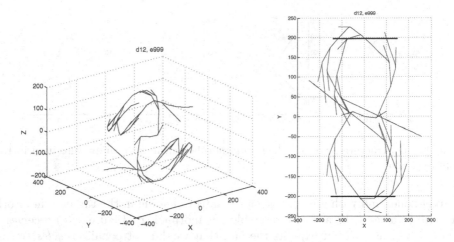

Fig. 7. The evolved structure that obeys our predefined function requirements. Left: 3D view. Right: 2D XY projection.

sampling of parameters from empirical distributions to generate virtual neurons (so called *a priori limitation strategy* [28]). Consequently, the only candidates morphologies they can generate are morphologies that statistically fit with biological neurons. To put it differently, if they optimize for a specific function, the evolved morphology will reflect sampled data, and will thus resemble known neuronal morphologies. In this light, is it less surprising that they had a good matching between their generated virtual neuron and the Nucleus Laminaris neuron which function they tried to approximate. We argue that the a priori limitation strategy is rather constraining the search for a mapping between morphology and function. Contrastingly, we use an algorithm that considers all 3D structures as candidates, and are thus unbiased towards specific solutions (i.e., morphologies) [28]. We consider this unbiased nature of our generation algorithm an advantage in the study of the morphology-function relationship.

As we just pointed out the advantages of our approach we also observe limitations of the synthetic approach.By using a descriptive model to generate morphologies, we do not include any underlying mechanisms of neuronal development. We are aware that these biological constraints heavily determine the final shape of the neuron. For instance, a parsimonious principle will be obeyed in the brain as there is only a limited amount of neuronal building blocks like micro-tubules and F-actin [21]. Energy consumption, and genetic borders are also known to limit the size of neurons [21]. Nevertheless, we can add extra - superficial - requirements to the fitness function to capture some underlying biological principles. Figure 8 illustrates four XY projections of structures that were found when we only assessed a structure's fitness on the basis of morphological properties representing morphological properties of NL neurons. The rationale was to evolve small (i.e., short total length) structures displaying basic properties of neurons (i.e., only bifurcations) with as many segments as possible in the

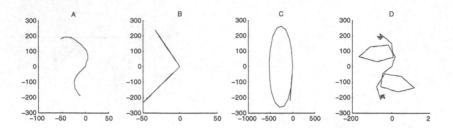

Fig. 8. 3D Structures evolved when *only* morphological constraints were used in the optimization

target zones and with a preference for dendrites ending in the target zone. Both Figure 8 (a) and (b) actually resemble NL morphology. Figure 8 (c) receives a high fitness as well but exploits the fact that we did not penalize *cycles* (recall these are projection in XY; only trees can be generated with EVOL-NEURON) and does not resemble neuronal morphology. Finally, Figure 8 (d) illustrates the strength of EC: this structure receives a high fitness as it has a great number of dendrites ending in the target zones (the two "balls" around Y 200 and -200). In the future, we want to combine a functional assessment with a morphological assessment to evolve structures that fulfill a specific function while resembling plausible neuronal morphology.

We can conclude that EVOL-NEURON (in combination with a physiological simulator) is a useful tool to investigate the neuronal morphology-function relationship. It allows the exploration of large parameter spaces, both morphological and functional. As a first result, we showed the non-triviality of neuronal morphologies. In future studies we will incorporate more biological constraints to advance us in the understanding of the relation between morphology and function.

Acknowledgements

The research reported here is part of the Interactive Collaborative Information Systems (ICIS) project, supported by the Dutch Ministry of Economic Affairs, grant nr: BSIK03024. The authors wish to thank Dr. Klaus Stiefel for discussions on this topic.

References

1. Hagai Agmon-Snir, Catherine E. Carr, and John Rinzel. The role of dendrites in auditory coincidence detection. *Nature*, 393:268–272, 1998.
2. Giorgio A. Ascoli. Mobilizing the base of neuroscience data: the case of neuronal morphologies. *Nature Neuroscience Reviews*, 7:318–324, 2006.
3. Giorgio A. Ascoli and Jeffrey L. Krichmar. L-Neuron: a modeling tool for the efficient generation and parsimonious description of dendritic morphology. *Neurocomputing*, 32-33:1003–1011, 2000.

4. Giorgio A. Ascoli, Jeffrey L. Krichmar, Slawomir J. Nasuto, and Stephen L. Senft. Generation, description and storage of dendritic morphology. *Phil. Trans. R. Soc. Lond. B*, 356:1131–1145, 2001.
5. Giorgio A. Ascoli, Jeffrey L. Krichmar, Ruggero Scorcioni, Slawomir J. Nasuto, and Stephen L. Senft. Computer generation and quantitative morphometric analysis of virtual neurons. *Anat. Embryol.*, 204:283–301, 2001.
6. N. Carnevale and M. Hines. *The NEURON book*. Cambridge University Press, 2006.
7. Christos Consoulas, Carsten Duch, Ronald J. Bayline, and Richard B. Levine. Behavioral transformations during metamorphosis: remodeling of neural and motor systems. *Brain research bulletin*, 53 (5):571–583, 2000.
8. Carsten Duch and R.B. Levine. Remodeling of membrane properties and dendritic architecture accompanies the postembryonic conversion of a slow into a fast motorneuron. *J. NeuroScience*, 20(18):6950–6961, 2000.
9. J.P. Eberhard, A. Wanner, and G. Wittum. NeuGen: a toold for the generation of realistic morphology of cortical neurons and neural networks in 3d. *Neurocomputing*, XX:in press, 2006.
10. John Holland. *Adaptation in Natural and Artificial Systems*. University of Michigan Press, 1975.
11. William L. Kath. Computational Modelling of Dendrites. *J. Neurobiol*, 64:91–99, 2005.
12. Christif Koch and Idan Segev. The role of single neurons in information processing. *Nature Neuroscience*, 3:1171–1177, 2000.
13. John Koza. *Genetic programming: On the programming of computers by means of Natural Selection*. MIT Press, Cambridge, 1992.
14. Jeffrey L. Krichmar, Slawomir J. Nasuto, Ruggero Scorcioni, and Stuart D. Washington. Effects of dendritic morphology on CA3 pyramidal cell electrophysiology: a simulation study. *Brain Res.*, 941:11–28, 2002.
15. Jyh-Ming Lien, Marco Morales, and Nacy M. Amato. Neuron PRM: A Framework for Constructing Cortical Networks. *Neurocomputing*, 52–54:191–197, 2003.
16. A. Lindenmayer. Mathematical models for cellular interactions in development i & ii. *Journal of Theoretical Biology*, 18:280–315, 1968.
17. Michael London and Michael Häusser. Dendritic computation. *Annu. Rev. Neurosci.*, 25:5003–532, 2005.
18. Zachary F. Mainen and Terrence J. Sejnowski. Influence of dendritic structure on firing pattern in model neocortical neurons. *Nature*, 382:363–366, 1996.
19. M. Mitchell. *An Introduction to Genetic Algorithms*. Cambridge, MA: MIT Press, 1996.
20. Przemyslaw Prusinkiewicz and Aristid Lindenmayer. *The algorithmic beauty of plants*. Springer-Verlag, 1990.
21. E.K. Scott and L. Luo. How do dendrites take their shape? *Nature (neuroscience)*, 4:(4)359–365, 2001.
22. Idan Segev. Sound grounds for computing dendrites. *Nature*, 393:207–208, 1998.
23. Idan Segev and Michael London. Untangling dendrites with quantitative models. *Science*, 290:744–749, 2000.
24. Armen Stepanyants and Dmitri B. Chklovskii. Neurogeometry and potential synaptic connectivity. *Trends in Neurosciences*, 28(7):387–394, 2005.
25. Volker Steuber, Erik De Schutter, and Dieter Jaeger. Passive model of neurons in the deep cerebellar nuclei: the effect of reconstruction errors. *Neurocomputing*, 58–60:563–568, 2004.

26. Klaus M. Stiefel and Terrence J. Sejnowski. Mapping function onto neuronal morphology. *j. Neurophysiol*, XX:XX, 2006 (in press).
27. Benjamin Torben-Nielsen, Karl Tuyls, and Eric O. Postma. Shaping realistic neuronal morphologies: An evolutionary computation method. In *International Joint Conference on Neural Networks (IJCNN2006), Vancouver, Canada*, 2006.
28. Benjamin Torben-Nielsen, Karl Tuyls, and Eric O. Postma. EvOL-Neuron: Neuronal Morphology Generation. submitted.
29. Arjen van Ooyen, Jacob Diujnhouwer, Michiel W. H. Remme, and Jaap van Pelt. The effect of dendritic topology on firing patterns in model neurons. *Network: Computation in Neural Systems*, 13:311–325, 2002.

Analyzing Stigmergetic Algorithms Through Automata Games

Peter Vrancx*, Katja Verbeeck, and Ann Nowé

Computational Modeling Lab,
Vrije Universiteit Brussel
{pvrancx, kaverbee, ann.nowe}@vub.ac.be

Abstract. Stigmergy describes a class of mechanisms that mediate animal to animal interaction through the environment. Recently this concept has proved interesting for use in multi-agent systems, as it provides a simple framework for agent interaction and coordination. However, determining the global system behavior that will arise from local stigmergetic interactions is a complex problem. In this paper stigmergetic mechanisms are modeled using simple reinforcement learners, called learning automata.We show that using automata to model stigmergy, the learning problem can be asymptotically approximated by an automata game. Existing convergence results for automata games enables us to understand these stigmergetic methods and predict their global behavior. A simple multi-pheromone example is described and analyzed through its corresponding automata game.

1 Introduction

The concept of stigmergy [13] was first introduced by entomologist Paul Grassé [5] to describe indirect interactions between termites building a nest. Generally stigmergy is defined as a class of mechanisms that mediate animal to animal interaction through the environment. The idea behind stigmergy is that individuals coordinate their actions by locally modifying the environment rather than by direct interaction. The changed environmental situation caused by one animal, will stimulate others to perform certain actions. This concept has been used to explain the coordinated behavior of social insects such as termites, ants and bees.

Recently the notion of stigmergy has gained interest in the domains of multi-agent systems and agent based computing [15,8,14,10]. Algorithms such as Ant Colony Optimization (ACO) [4] model aspects of social insect behavior to coordinate agent behavior and cooperation. The concept of stigmergy is promising in this context, as it provides a relatively simple framework for agent communication and coordination. One of the main problems that arises, however, is the difficulty of determining the global system behavior that will arise from local stigmergetic interactions.

---

* Funded by a Ph.D grant of the Institute for the Promotion of Innovation through Science and Technology in Flanders (IWT Vlaanderen).

K. Tuyls et al. (Eds.): KDECB 2006, LNBI 4366, pp. 145–156, 2007.

In this paper we propose to model stigmergetic communication by networks of learning automata (LA)[7]. One of the main advantages of LA is the existence of convergence proofs for many possible interaction schemes. Using LA as a model for stigmergy allows us to transfer some of these insights to stigmergetic algorithms. A principal contribution of LA theory is that a set of decentralized learning automata is able to control a finite Markov Chain with unknown transition probabilities and rewards [16]. The similarity between this approach and simple Ant Colony Optimization algorithms was already shown in [15]. Here we will use and extend this result to more general stigmergetic methods.

We show that a modification of the interconnected network of learning automata used in [16] is also able to model multi-pheromone ant algorithms. We show how this new model can still be asymptotically approached by automata games as was the case in [16,15]. The large advantage of this is that existing automata theory[11] will allow us to analyze stigmergetic algorithms in terms of their global behaviour. In case of simple single type pheromone updates as in the S-ACO model, convergence is guaranteed to a global optimum as long as the underlying network satisfies some ergodic assumption. In case of multi-type pheromone updates convergence is still assured to possibly suboptimal attractor points.

The remainder of this paper is organized as follows. We first introduce stigmergetic algorithms. We focus on 2 methods; a simple algorithm that uses stigmergy to coordinate agent behavior and a more complex multi-type method that allows for agent competition. We then define learning automata and automata games. In section 4, we explain how a set of interconnected LA can model the stigmergetic algorithms given in section 2. Learning automata theory is then used to analyze the corresponding automata games. Finally we demonstrate our result on a simple multi-pheromone example.

2   Stigmergetic Algorithms

Several different approaches have been proposed to apply stigmergy to multi-agent systems. A commonly used method is to let agents communicate by using artificial pheromones. Agents can observe and alter local pheromone values which guide action selection. This system has been used in optimization [3] and manufacturing control [14,2], among others. An example of this type of algorithm is given in the next subsection.

Other algorithms are based on termite and wasp nest building behavior [12] or ant brood sorting [1,6]. In these systems an individual's actions (e.g. building, depositing dirt, picking up brood) modify the local environment and cause reactions of other workers (i.e building more, moving brood, etc.).

In most of the algorithms based on the systems mentioned above a set of common elements can be isolated:

- The agent environment is subdivided in a number of discrete locations, which agents can visit.
- Each location contains a local state that can be accessed and updated by agents visiting that location.

- Agents can perform actions in the location they visit. The probability of an action is determined by an agent based on the local state.
- An interconnection scheme between locations is defined, allowing agents to travel between locations.

We will try to accommodate these recurring elements using our learning automata framework.

2.1 A Simple Stigmergetic Algorithm

In this section we describe an algorithm called S-ACO (Simple Ant Colony Optimization). The algorithm was proposed in [4] to study the basic properties of ant colony optimization algorithms. It contains all the basic mechanisms used in algorithms that employ artificial pheromones.

The goal of the S-ACO algorithm is to find the minimum cost path between 2 nodes in a weighted graph. The algorithm uses a colony of very simple ant-like agents. These agents travel through the graph starting from the source node, until they reach the destination node. In all nodes a pheromone value is associated with the outgoing edges. When an agent arrives in the node it reads these values and uses them to assign a probability to each edge. This probability is used to choose an edge to follow to the next node. When all agents have reached the goal state they return to the source node and the pheromone value τ_{ij} on each edge ij is updated using the following formula:

$$\tau_{ij} \leftarrow \rho\tau_{ij} + \Delta\tau_{ij} \tag{1}$$

The pheromone update described above consists of two parts. First the pheromones on the edges are multiplied with a factor $\rho \in [0, 1]$. This simulates pheromone evaporation and prevents the pheromones from increasing without bound. After evaporation a new amount of pheromone $\Delta\tau$ is added. The amount of new pheromones is determined by the paths found by the ant agents. Edges that are used in lower cost paths receive more pheromones than those used in high cost paths. By using this system the ant agents are able to coordinate their behavior until all agents follow the same shortest path.

2.2 Noncooperative Algorithms

The algorithm described in the previous section is one of the simplest stigmergetic algorithms possible. All agents have the same goal and they alter their environment (by leaving pheromones) to share information and coordinate their actions. The agents do not influence each other's reward, however. Each agent builds it own path, and the feedback it gets is based solely on the cost of this path. The paths followed by other agents do not influence this feedback.

In this paper we also examine more complex problems where agents can have different goals and directly influence each other's reward. Examples of these algorithms can be found in multi-pheromone algorithms. These systems use not one, but several pheromone gradients to guide agents. One such system was proposed

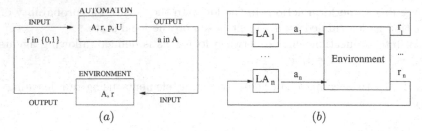

Fig. 1. (a) A Single Learning Automata - Environment pair. (b) Multiple automata sharing an environment in an Automata Game.

in[9] to let different colonies of agents find disjoint paths in a network. Another multi-pheromone system was proposed in [10]. Here the different pheromones guide agents to different locations.

Depending on their current goal, different agents can prefer different actions in the same location. Furthermore it is possible that agent goals conflict and agents influence each other's rewards. In the disjoint paths problem mentioned above for instance, different agents compete for the use of network resources and an agent's reward is determined by how many other agents use the same network links.

3   Learning Automata

Learning Automata are simple reinforcement learners originally introduced to study human behavior. The objective of an automaton is to learn an optimal action, based on past experience. Formally the automaton is described by a quadruple $\{A, r, p, U\}$ where $A = \{a_1, \ldots, a_n\}$ is the set of possible actions the automaton can perform, p is the probability distribution over these actions, r is a random variable between 0 and 1 representing the evironmental response, and U is a learning scheme used to update p.

A single automaton is connected in a feedback loop with its environment. Actions chosen by the automaton are given as input to the environment and the environmental response to this action serves as input to the automaton. This situation is represented in Figure 1(a).

Several automaton update schemes with different properties have been studied. Important examples of linear update schemes are linear reward-penalty, linear reward-inaction and linear reward-ϵ-penalty. The philosophy of these schemes is essentially to increase the probability of an action when it results in a success and to decrease it when the response is a failure. The general algorithm is given by:

$$p_i(t+1) = p_i(t) + \alpha_{reward}\, r(t)(1 - p_i(t))$$
$$- \alpha_{penalty}(1 - r(t))p_i(t) \tag{2}$$
$$\text{if } a_i \text{ is the action taken at time step } t$$

$$p_j(t+1) = p_j(t) - \alpha_{reward} \, r(t)p_j(t)$$
$$+\alpha_{penalty}(1 - r(t))[(l-1)^{-1} - p_j(t)] \qquad (3)$$
$$\text{if } a_j \neq a_i$$

with l the number of actions of the action set A. The constants α_{reward} and $\alpha_{penalty}$ are the reward and penalty parameters respectively. When $\alpha_{reward} = \alpha_{penalty}$ the algorithm is referred to as linear reward-penalty (L_{R-P}), when $\alpha_{penalty} = 0$ it is referred to as linear reward-inaction (L_{R-I}) and when $\alpha_{penalty}$ is small compared to α_{reward} it is called linear reward-ϵ-penalty ($L_{R-\epsilon P}$).

3.1   Automata Games

Automata games, [7,11] were introduced to see if learning automata could be interconnected so as to exhibit group behavior that is attractive for either modeling or controlling complex systems.

As visualized in Figure 1(b), a play $a(t) = (a_1(t)\ldots a_n(t))$ of n automata is a set of strategies chosen by the automata at stage t. Correspondingly, the outcome is now a vector $r(t) = (r_1(t)\ldots r_n(t))$. At every instance, all automata update their probability distributions based on the responses of the environment. Each automaton participating in the game operates without information concerning the number of participants, their strategies, their payoffs or actions. The asymptotic behaviour of a learning automata game is already studied well. In zero-sum games the L_{R-I} scheme converges to the equilibrium point if it exist in pure strategies, while the $L_{R-\epsilon P}$ scheme can approach a mixed equilibrium arbitrarily close [7,11]. In general non zero sum games [7,11] it is shown that when the automata use a L_{R-I} scheme and the game is such that a unique pure equilibrium point exists, convergence is guaranteed. In cases where the game matrix has more than one pure equilibrium, which equilibrium is found depends on the initial conditions.

4   Analyzing Stigmergy Through Automata Games

The use of learning automata as a model for stigmergetic communication was first introduced in [15]. In this paper a LA model for S-ACO type algorithms was introduced. One limitation of this approach is that all agents must share the same goal, to avoid that the LAs are updated using conflicting responses. In this paper we aim to generalize this idea by allowing agents to have different goals. This could for instance be used to model the multiple pheromones systems descibed in section 2.2.

4.1   A Model for S-ACO

The idea behind the LA model of [15] is to move decision making from the agents to the local environment states. Each local state contains one learning automata. When an agent visits a location it activates the learning automaton that resides in that location. This automaton then decides the action the agent should take in that location. Transition to the next location triggers an automaton from that location to become active and take some action.

Agents themselves can then be viewed as dummy mobile agents, that walk around in the graph of interconnected locations, make local LA active and bring information so that the LA involved can update their local state. Multiple agents can collaborate by using the same automata. The learning automaton LA_i active in location i is not informed of the one-step reward $r_j^i(k)$ resulting from its action a_k, leading to state j. When the agent visits location i again, LA_i receives two pieces of data: 1) the cumulative reward received by the agent up to the current time step and 2) the current global time. From these, LA_i computes the incremental reward generated since the last visit and the corresponding elapsed global time. The environment response or the input to LA_i is then taken to be:

$$r^i(n_i + 1) = \frac{R_k^i(n_i + 1)}{\eta_k^i(n_i + 1)} \tag{4}$$

where $R_k^i(n_i + 1)$ is the cumulative total reward generated for action a_k in state i and $\eta_k^i(n_i + 1)$ the cumulative total time elapsed[1].

When comparing the update pheromone update rule given in Equation 1 with the update scheme of the interconnected LA model, see Equations 2 and 3, the commonalities are obvious. Indeed, the trail update rule of Equation 1 is actually a reward-penalty (L_{R-P}) update. The pheromone trail in S-ACO is updated with an amount which depends on the total length or cost of the path. So depending on the quality of the path visited, the pheromone trail is rewarded or penalized. In fact, the dummy agents play the role of the ants here.

The LA model described above is based on the LA algorithm introduced by Wheeler and Narendra [16] to solve Markov Decision Problems. The only difference is that here multiple dummy agents co-exist, so that multiple automata can be activated at the same time. In [15] it is shown that the system is approached by a limiting automata game just as is the case for the original model in [16] when the following assumption is fulfilled: The Markov chain corresponding to each pure policy α is ergodic[2] This actually means that the agents should continue to visit each location. So there are no transient states and for each pure policy α a limiting distribution $\pi(\alpha) = (\pi_1(\alpha) \ldots \pi_N(\alpha))$ exists, with N the number of locations in the environment. Using these limiting distributions we can formulate the expected reward per step for each policy as follows:

$$J(\alpha) = \sum_{i=1}^{N} \pi_i(\alpha) \sum_{j=1}^{N} T_j^i(\alpha) r_j^i(\alpha) \tag{5}$$

---

[1] The one step reward is normalized so that r stays in $[0, 1]$.

[2] Note that the policies we consider, are limited to stationary, nonrandomized policies. Under the assumption that the Markov chain corresponding to each policy α is ergodic, it can be shown that the best strategy in any state is a pure strategy, independent of the time at which the state is occupied [16].

where $T_j^i(\alpha)$ and reward $r_j^i(\alpha)$ are respectively the transition probability to state j and immediate reward when action α is chosen in state i.

This also allows us to write down the limiting automata game as follows: each learning automata is assumed to be a player and each joint action or play of the game corresponds with a pure policy α. The resulting payoff is given by $J(\alpha)$ defined in Equation 5. For the model described here, the limiting automata game is proved to have a unique equilibrium [16]. As described in section 3.1, L_{R-I} automata are able to converge to this optimal point. In the extended model we will loose the guarantee to find optimal equilibria or policies, but we are still able to show convergence to a local optimum.

4.2   A Model for Noncooperative Stigmergy

One limitation of the above model is that all agents must share the same goal, to avoid that the LA are updated using conflicting responses. We now extend the model to allow agents to have different goals. Therefore in the extended model, each location contains one or more learning automata, corresponding to different goals agents are trying to achieve. When an agent visits a location it activates the learning automaton corresponding to its goal. For each agent, one LA in its current location is active at each time step and the transition to the next location triggers an automaton from that location to become active and take some action. The update of the LA is completely the same as before. Using the cumulative reward and the current global time that the agent brings on its next visit, the environment response β given in Equation 4, is computed and used in the update scheme of Equations 2 and 3.

To see how this model can still be approached by a limiting game, we redefine the notion of a state, i.e. locations and states are no longer the same. Since agents have competing goals, we are going to include the other agents in the defintion of our state space. For every combination of all agent's locations we get another joint state. When one or more agents move to another location, a transition to another joint state takes place. In this joint state space we can just as above write down the limiting automata game. This is demonstrated on a simple grid world problem in the next section.

4.3   Examples

We first demonstrate our analysis on the small grid world problem shown in Figure 2(a). The game consists of only two grid locations $L1$ and $L2$. Two agents A and B try to coordinate their behavior in order to receive the maximum reward. Each time step both agents take one of 2 possible actions. If an agent chooses action 0 it stays in the same location, if it chooses action 1 it moves to the other grid location. The transitions in the grid are stochastic. An agent has a probability of 0.9 to arrive in the chosen location and a probability of 0.1 to arrive in the other one. The reward function for both agents is as follows:

$$r(t) = \begin{cases} 1 & \text{if agent 1 is in location 1 and agent 2 is in location 2} \\ 0.01 & \text{else} \end{cases}$$

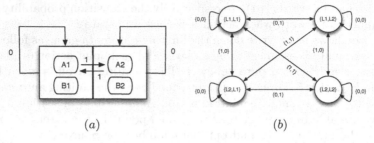

(a)   (b)

Fig. 2. (a)The grid-world game with 2 grid locations and 2 non-mobile LA in every location. (b) Markov game representation of the same game.

We apply the LA model described in the previous section to this grid game. Each agent has a learning automaton in both locations. This learning automaton decides the action the corresponding agent takes in that state and is updated with the cumulative reward whenever the agent visits its location.

The game described above can be transformed to a Markov game by considering the product space of the locations and actions. A state in the Markov game consists of the locations of both agents, e.g. $S1 = \{L1, L1\}$ when both agents are in grid cel 1. The actions that can be taken to move between these states are the joint actions taken by all agents. Figure 2(b) represents the markov game corresponding to our grid world problem.

Because automata in the stigmergy framework are associated with the possible agent location, rather than the Markov game states, it is not possible to learn all possible policies. For instance the automaton $A1$, associated with location $L1$ is present state $S1 = \{L1, L1\}$ as well as state $S2 = \{L1, L2\}$. Therefore it is not possible for agent A to learn a different action in state $S1$ and $S2$. This corresponds to an agent associating actions with locations, without modeling the other agents.

To analyze the grid world problem we determine the automata game corresponding to the problem. As stated in Section 4.1 , we can calculate the expected reward for each policy in the Markov game, provided that the corresponding markov game policy corre-sponds ... Here the chain is ergodic and we can calculate transition probabilities from the data of the grid world problem. For instance in location $\{L1, L1\}$ with joint action $\{0, 0\}$ chosen, the probability to stay in state $\{L1, L1\}$ is 0.81. The probabilities corresponding with moves to states $\{L1, L2\}$, $\{L2, L1\}$ and $\{L2, L2\}$ are $0.09, 0.09$ and 0.01 respectively. The transition prob-abilities for all states and joint action pairs can be calulated this way. With the transition probabilities and the rewards known, we can use Equation 5 to calculate the expected reward.

The complete game is shown in Table 1. Columns 1 through 4 show the play selected by the automata, columns 5 to 8 show the Markov game policy corre-sponding to this play chosen. This policy maps one joint action to each state in the Markov game. Finally column 9 shows the expected reward for the given

Table 1. Possible LA actions with the corresponding policies in the Markov game and their expected reward. The unique equilibrium reward is indicated in bold.

Agent A		Agent B		Policy				Expected reward
A1	A2	B1	B2	$\{L1, L1\}$	$\{L1, L2\}$	$\{L2, L1\}$	$\{L2, L2\}$	
0	0	0	0	$\{0,0\}$	$\{0,0\}$	$\{0,0\}$	$\{0,0\}$	0.2575
0	0	0	1	$\{0,0\}$	$\{0,1\}$	$\{0,0\}$	$\{0,1\}$	0.0595
0	0	1	0	$\{0,1\}$	$\{0,0\}$	$\{0,1\}$	$\{0,0\}$	0.4555
0	0	1	1	$\{0,1\}$	$\{0,1\}$	$\{0,1\}$	$\{0,1\}$	0.2575
0	1	0	0	$\{0,0\}$	$\{0,0\}$	$\{1,0\}$	$\{1,0\}$	0.4555
0	1	0	1	$\{0,0\}$	$\{0,1\}$	$\{1,0\}$	$\{1,1\}$	0.0991
0	1	1	0	$\{0,1\}$	$\{0,0\}$	$\{1,1\}$	$\{1,0\}$	**0.8119**
0	1	1	1	$\{0,1\}$	$\{0,1\}$	$\{1,1\}$	$\{1,1\}$	0.4555
1	0	0	0	$\{1,0\}$	$\{1,0\}$	$\{0,0\}$	$\{0,0\}$	0.0595
1	0	0	1	$\{1,0\}$	$\{1,1\}$	$\{0,0\}$	$\{0,1\}$	0.0199
1	0	1	0	$\{1,1\}$	$\{1,0\}$	$\{0,1\}$	$\{0,0\}$	0.0991
1	0	1	1	$\{1,1\}$	$\{1,1\}$	$\{0,1\}$	$\{0,1\}$	0.0595
1	1	0	0	$\{1,0\}$	$\{1,0\}$	$\{1,0\}$	$\{1,0\}$	0.2575
1	1	0	1	$\{1,0\}$	$\{1,1\}$	$\{1,0\}$	$\{1,1\}$	0.0595
1	1	1	0	$\{1,1\}$	$\{1,0\}$	$\{1,1\}$	$\{1,0\}$	0.4555
1	1	1	1	$\{1,1\}$	$\{1,1\}$	$\{1,1\}$	$\{1,1\}$	0.2575

policy. If all automata in the grid use the *Reward-Inaction* learning scheme, the automata are guaranteed to converge to an equilibrium. In this case there is only one, optimal equilibrium, so the automata will learn the optimal action.

Figure 3 shows the probability of action 0 for the automata in the game. It can be seen that the automata $A1$ and $B2$ converge to action 0, while $A2$ and $B1$ converge to action 1. This corresponds to the equilibrium policy indicated in Table 1.

We now demonstrate our approach experimentally on a larger grid world game, shown in Figure 4 . This game was inspired by the disjoint paths algorithms mentioned in Section 2.2. The game consists of a 3×3 grid and four agents who are trying to reach their goal state. The agents start in the corners of the grid, i.e. squares 0, 2, 6 and 9. Agents 1 and 2 are trying to reach goal B, while agents 3 and 4 try to reach goal A.

If multiple players attempt to move to the same square, different from their goal state, the moves fail and the agents are bounced back to their previous states. The agents' goal is to find a shortest path to the goal state without interfering with each other.

The reward function used here for each agent i is the following:

$$r_i(t) = \begin{cases} 1 & \text{if agent i reached the goal state} \\ 0.001 & \text{if both agents reached a different non-goal state} \\ 0 & \text{if both agents reached the same non-goal state} \end{cases}$$

When all agents have reached the goal, they are put back in their initial states.

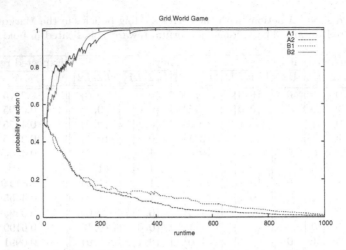

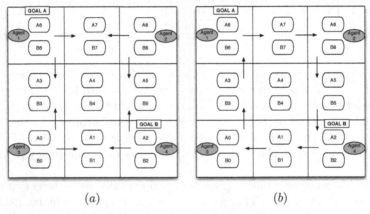

Fig. 4. (a) The grid-world game with 4 mobile agents in their initial state and 2 non-mobile LA in every grid location. (b) Solution found by agents using the LA algorithm.

Using the LA model we place 2 learning automata in each location. These correspond to the goal locations A and B. Agents 1 and 2 use the automata corresponding to goal B, while agents 3 and 4 share the LA corresponding to goal B. Each automaton has as many actions as its location has neighbors. When an agent leaves a grid location, the action chosen by the automaton determines the neighbor location it travels to. Transitions in the grid are again stochastic, with agent having a probability of 0.9 of arriving in the chosen location and a 0.1 probability of arriving in a randomly chosen neighbor location.

Figure 4(b) shows the result found by the LA model with all automata using the L_{R-I} update scheme. This result is an equilibrium solution corresponding

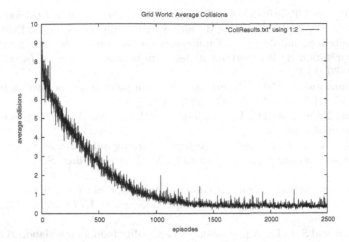

Fig. 5. The average number of collision in each iteration of the gridworld game using $\alpha_{penalty} = 0.06$. Average taken over 100 game runs.

to non-conflicting paths to the goal locations. As can be seen in Figure 5 the algorithm minimizes the number of collisions over time. Eliminating collisions entirely is not possible in this case, due to the stochastic transitions.

5   Conclusion

In this paper we model 2 types of stigmergetic mechanisms as interconnected models of learning automata. These models can be asymptotically approached by an automata game as was the case in [16,15]. The large advantage of this is that existing automata theory [11] allows us to analyze stigmergetic algorithms in terms of their global behaviour. In case of simple single type pheromone updates as in the S-ACO model, convergence is guarenteed to a global optimum as long as the underlying network satisfies some ergodic assumption. In case of noncooperative updates convergence is still assured to local attractor points. As far as we know the analysis method we present here is one the first formal analysis tool for stigmergetic mechanisms. Of course, further study is necessary to see whether the approximations are still good enough in larger networks.

References

1. R. Beckers, OE Holland, and J.L. Deneubourg. From local actions to global tasks: Stigmergy and collective robotics. *Artificial Life IV*, 181:189, 1994.
2. S. Brueckner. *Return from the Ant: Synthetic Ecosystems for Manufacturing Control.* PhD thesis, PhD Dissertation, Humboldt-Universitat Berlin, Germany (2000), 1999.
3. M. Dorigo, E. Bonabeau, and G. Theraulaz. Ant algorithms and stigmergy. *FUTURE GENER COMPUT SYST*, 16(8):851–871, 2000.

4. M. Dorigo and T. Stützle. *Ant Colony Optimization*. Bradford Books, 2004.
5. P.P. Grassé. La reconstruction du nid et les coordinations interindividuelles chez Bellicositermes natalensis et Cubitermes sp. la théorie de la stigmergie: Essai d'interprétation du comportement des termites constructeurs. *Insectes Sociaux*, 6(1):41–80, 1959.
6. O. Holland and C. Melhuish. Stigmergy, self-organization, and sorting in collective robotics. *Artificial Life*, 5(2):173–202, 1999.
7. K. Narendra and M. Thathachar. *Learning Automata: An Introduction*. Prentice-Hall International, Inc, 1989.
8. A. Nowé, K. Verbeeck, and M. Peeters. Learning automata as a basis for multi-agent reinforcement learning. *Lecture Notes in Computer Science*, 3898:71–85, 2006.
9. A. Nowé, K. Verbeeck, and P. Vrancx. Multi-type ant colony: The edge disjoint paths problem. *Lecture Notes in Computer Science: ANTS 2004*, 3172:202 – 213, 2004.
10. L. Panait and S. Luke. A pheromone-based utility model for collaborative foraging. *Autonomous Agents and Multiagent Systems, 2004. AAMAS 2004. Proceedings of the Third International Joint Conference on*, pages 36–43, 2004.
11. M.A.L. Thathachar and P.S. Sastry. *Networks of Learning Automata: Techniques for Online Stochastic Optimization*. Kluwer Academic Publishers, 2004.
12. G. Theraulaz and E. Bonabeau. Modelling the collective building of complex architectures in social insects with lattice swarms. *Journal of Theoretical Biology*, 177(4):381–400, 1995.
13. G. Theraulaz and E. Bonabeau. A brief history of stigmergy. *Artificial Life*, 5(2):97–116, 1999.
14. P. Valckenears and M. Kollingbaum. Multi-agent coordination and control using stigmergy applied to manufacturing control. *Mutli-agents systems and applications*, pages 317–334, 2001.
15. K. Verbeeck and A. Nowe. Colonies of learning automata. *Systems, Man and Cybernetics, Part B, IEEE Transactions on*, 32(6):772–780, 2002.
16. R. Wheeler Jr and K. Narendra. Decentralized learning in finite Markov chains. *Automatic Control, IEEE Transactions on*, 31(6):519–526, 1986.

The Identification of Dynamic Gene-Protein Networks

Ronald L. Westra[1], Goele Hollanders[2], Geert Jan Bex[2],
Marc Gyssens[2], and Karl Tuyls[1]

[1] Department of Mathematics and Computer Science,
Maastricht University and Transnational University of Limburg,
Maastricht, The Netherlands
[2] Department of Mathematics, Physics, and Computer Science,
Hasselt University and Transnational University of Limburg,
Hasselt, Belgium
westra@math.unimaas.nl

Abstract. In this study we will focus on piecewise linear state space models for gene-protein interaction networks. We will follow the dynamical systems approach with special interest for partitioned state spaces. From the observation that the dynamics in natural systems tends to punctuated equilibria, we will focus on piecewise linear models and sparse and hierarchic interactions, as, for instance, described by Glass, Kauffman, and de Jong. Next, the paper is concerned with the identification (also known as reverse engineering and reconstruction) of dynamic genetic networks from microarray data. We will describe exact and robust methods for computing the interaction matrix in the special case of piecewise linear models with sparse and hierarchic interactions from partial observations. Finally, we will analyze and evaluate this approach with regard to its performance and robustness towards intrinsic and extrinsic noise.

Keywords: piecewise linear model, robust identification, hierarchical networks, microarrays, gene regulatory networks.

1 Introduction and Problem Statement

This paper is concerned with the identification of dynamic gene-protein interaction networks with intrinsic and extrinsic noise from empirical data, such as a set of microarray time series. Prerequisite for the successful reconstruction of these networks is the way in which the dynamics of their interactions is modeled. In the past few decades, a number of different formalisms for modeling the interactions amongst genes and proteins have been presented. Some authors focus on specific detailed processes such as the circadian rhythms in *Drosophila* and *Neurospora* [9,10], or the cell cycle in *Schizosaccharomyces* (fission yeast) [12]. Others try to provide a general platform for modeling the interactions between genes and proteins. For a thorough overview, consult de Jong in [2], Bower in [1], and others [5,11]. We will focus on dynamical models, and not discuss static models where the relations between genes are considered fixed in time. A dynamical model can be described using continuous time, or discrete events (or time). Given the discrete nature of the data we have at our disposal to derive the models, a discrete event model seems most appropriate. In discrete event simulation models, the detailed

K. Tuyls et al. (Eds.): KDECB 2006, LNBI 4366, pp. 157–170, 2007.
© Springer-Verlag Berlin Heidelberg 2007

biochemical interactions are studied. Considering a large number of constituents, the approach aims to derive macroscopic quantities. More information on discrete event modeling can be found in [1].

2 Modeling Dynamic Gene-Protein Interactions as a Piecewise Linear System

A frequent approach to modeling the dynamical interactions amongst genes and proteins is to consider them as biochemical reactions, and thus represent them as 'rate equations', i.e. as a set of differential equations, expressing the time derivative of the concentration of each constituent of the reaction as some rational function of the concentrations of all the constituents involved. In case of biochemical interactions between genes and proteins, the applicability of the concept of rate equations is valid only for genes with sufficient high transcription rates. This is confirmed by recent experimental findings by Swain and Elowitz [4,16,18,19]. A practical problem is that the precise details of most reactions are unknown, and therefore cannot be modeled as rate equations. This could be compensated by a well-defined parametrized generic form of the interactions, in which the parameters can be estimated from sufficient empirical data. A generic form based on rational positive functions is proposed by J. van Schuppen [21]. However, in the few cases where parts of such interaction networks have been described from experimental analysis, like the circadian rhythms in certain amoeba [9], or the cell cycle in fission yeast [12], it is clear that such forms have a too extensive syntax to be of any practical use.

Let us for now ignore these problems, and consider the dynamics of gene-RNA-protein networks. When we assume a stochastic differential equation as a model for the dynamics of the interaction network, the relation can be expressed as

$$\dot{x} = f(x, u|\theta) + \xi(t) \tag{1}$$

Here, $x(t)$, called the state-vector, denotes the N gene expressions and RNA/protein densities at time t—possibly involving higher order time derivatives; $u(t)$ denotes the P controlled inputs to the system, such as the timing and concentrations of toxic agents administered to the system observed; and $\xi(t)$ denotes a stochastic Gaussian white noise term. This expression involves a parameter vector θ that contains the coupling constants between gene expressions and protein densities. We can consider this system as being represented by the state vector $x(t)$ that wanders through the N-dimensional space of all possible configurations. In the formalism of dynamic systems theory, x will eventually enter an area of attraction, and become subject to the influence of an attractor. An attractor here can be a uniform convergent attractor, a limit cycle, or a 'strange attractor'. We can understand the entire space as being partitioned into cells, with each cell having an attractor or a repeller. Thus, the behavior of x can be described by motion through this collection of cells, swiftly moving through cells of repellers, until they enter the basin of attraction of an attractor. Under the effects of external agents via the vector $u(t)$ or by stochastic fluctuations via $\xi(t)$ they can leave this cell, and start wandering again, thereby repeating the process. Now, a vital assumption is that in each cell the behavior is governed by its specific (un)stable equilibrium point. In that case, it is possible to approximate the dynamics of Equation (1) in cell ℓ—for x near the ℓ-th equilibrium $x_{eq}^{(\ell)}$ and small u—(except the noise term) as:

$$\dot{x}(t) \approx \frac{\partial f(x_{eq}^{(\ell)}, u)}{\partial x}(x - x_{eq}^{(\ell)}) + \frac{\partial f(x_{eq}^{(\ell)}, u)}{\partial u}u \equiv A_\ell x(t) + B_\ell u(t) + c_\ell \qquad (2)$$

Thus, the qualitative behavioral dynamics of gene-protein interactions is characterized as predominantly linear behavior near the stable equilibria—called the steady states, interrupted by abrupt transitions where the system quickly relaxes to a new steady state, either externally induced or by process noise.

In biology, such behavior is frequently observed, as, for instance, in embryonic growth where the organism develops by transitions through a number of well-defined 'checkpoints'. Within each such checkpoint, the system is in relative equilibrium, see [20]. We will follow the view of *piecewise linear behavior* (PWL). This approach corresponds to the piecewise linear models introduced by Glass and Kauffman [8], and the qualitative piecewise linear models described by de Jong et al. [2,3].

3   Identification of Dynamic Networks Using *Piecewise Linear Models*

Next, we will be concerned with the identification (also known as *reverse engineering*) of piecewise linear gene regulatory systems from microarray data. We consider the case where time series of genome-wide expression data are available. The nature of our problem—few microarray experiments and lots of genes—implies that we are dealing with *poor data*, where the number of measurements is *a priori* insufficient to identify all parameters of the system. One standard approach to circumvent this problem is by dimension reduction through the clustering of related genes. A different perspective is offered by including some characteristics of the biological problem, such as the hierarchy and sparsity of the networks. The case of the identification of a *simple* linear system with sparse and hierarchic interactions is discussed by Peeters and Westra [14,23], and Yeung et al. [24]. In realistic situations, this model is too simple however. As was pointed out by Øyehaug et al. [13], such systems tend to behave in a switch-like manner, and they determine the switching timepoints using complex biological modeling. In contrast, we will determine the switching timepoints by identifying sparse *piecewise linear systems*. As a consequence, our focus is on modeling the subsystems between the switching points rather than on the dynamics of the switching points themselves, as, e.g., in Plahte et al. [15]. More concretely, our main aim is to obtain the local gene-gene interaction matrices A_ℓ, that directly relate to the graph of the gene regulatory network. Additionally, the matrices B_ℓ provide information on the coupling of genes to specific inputs.

3.1   General Dynamics of Switching Subsystems

In what follows, let us assume a dynamical input-output system Σ that switches irregularly between K linear time-invariant subsystems $\{\Sigma_1, \Sigma_2, \ldots, \Sigma_K\}$.

Let $S = \{s_1, s_2, \ldots, s_{K-1}\}$ denote the set of—unknown—switching times, i.e., the time instants $t = s_\ell$ when the system switches from subsystem Σ_ℓ, to $\Sigma_{\ell+1}$. Similarly as with the simple linear networks, we assume empirical data $X = (x[1], \ldots, x[M])$, $U = (u[1], \ldots, u[M])$, and $\dot{X} = (\dot{x}[1], \ldots, \dot{x}[M])$ at M sampling times $T = \{t_1, t_2, \ldots, t_M\}$, representing full observations of the N states and P inputs, and $x[k] \equiv x(t_k)$. The interval between two sample instants is denoted as $\tau_k = t_{k+1} - t_k$. Here, we assume that the system is sampled on regular time intervals, i.e., that the sample intervals are equal to τ. Within one subsystem Σ_ℓ, the effect of the inputs $u(t)$ is represented as a state-space system of first-order differential (for continuous time systems) or difference equations (for discrete time systems), using an internal vector $x(t)$ spanning the so-called subspace. In our case, this represents the observed gene expressions. In the case of continuous time and in the absence of noise, this system can be written as:

$$\dot{x}(t) = A_\ell x(t) + \tilde{B}_\ell \tilde{u}(t), \tag{3}$$

with $\tilde{B}_\ell = (B_\ell | - A_\ell e_\ell)$, $\tilde{u}^T = (u^T, 1)$, where e_ℓ indicates the equilibrium point of the ℓ-th subsystem and A_ℓ and B_ℓ refer to Equation (2). We will use this linear expression, and from here on drop the *tilde*. A general disadvantage is that the time evolution of the different genes, i.e., $x_v(t)$, $v = 1, \ldots, n$, will strongly correlate, thus obscuring their true relation. This can be avoided by using Equation (3) with time series of triplets $\xi[k] \equiv (x[k], u[k], \dot{x}[k])$ with a sufficient amount of statistically independent and varying inputs $u(t_k)$. Practically, this opens the way to combining distinct empirical sets. However, a practical disadvantage of Equation 3 is that the derivative $\dot{x}(t_k)$ can only be approximated from the measurements, such as $\dot{x}[k] \approx (x[k] - x[k-1])/(t_k - t_{k-1})$.

We furthermore assume that the system matrices in these equations are constant during intervals $[s_\ell, s_{\ell+1}[$, and abruptly change at the transition between the intervals at $t = s_{\ell+1}$. We assume that on the time scale τ, the system has relaxed to its new state. This means that we do not observe *mixed states*, which would severely complicate the problem of identification, e.g., see [22]. This is accomplished by defining *weights* $w_{k,\ell}$ as the degree to which observation k belongs to subsystem Σ_ℓ. If observation $\xi[k]$ belongs to system Σ_ℓ then $w_{k\ell} = 1$. Non-integer values in $[0,1]$ can be interpreted as the fuzzy membership of observation k to system Σ_ℓ. Since we assume that the subsystems $\{\Sigma_1, \Sigma_2, .., \Sigma_K\}$ act disjointly and subsequently, the result can be improved by matching the weights to a block function structure; i.e., $w_{kl} = 1$ for $t_k \in [s_l, s_{l+1}[$ and $w_{kl} = 0$ elsewhere. This may, however, introduce other problems, for instance if the same subsystem is revisited at different switching intervals. These considerations lead to the constraints C_{MK} on w:

$$C_{MK}(w) : \begin{cases} w_{1,1} = 1, w_{M,K} = 1, \\ \forall_{k,\ell} w_{k,\ell} \in [0,1], \\ \forall_k \sum_l w_{k,\ell} = 1, \\ \forall_\ell \sum_{k=1}^{M-1} |w_{k+1,1} - w_{k,1}| = 1, \\ \forall_\ell \sum_{k=1}^{M-1} |w_{k+1,\ell} - w_{k,\ell}| = 2, \\ \forall_\ell \sum_{k=1}^{M-1} |w_{k+1,K} - w_{k,K}| = 1. \end{cases} \tag{4}$$

3.2 Combining the System Matrices $\{A, B\}$ with the Subsystem Weightmatrix W

The assumption that the switching times between the linear subsystems are completely known suits various experimental conditions, as, for instance, when toxic agents are administered. In many biological situations, however, the exact timing between subsystems is *not* known, as during embryonic growth and in many metabolical processes.

When a sufficiently accurate record of estimates of the state derivatives $\dot{X}$ is available, we can simply rewrite this problem as a special case of the method described in the case of a simple linear problem as in [14]. In fact, by exploiting the data $\mathcal{D} = \{X, U, \dot{X}\}$, the problem can be stated as a linear equation in terms of new matrices H_1 and H_2 as

$$\dot{X} = H_1 X + H_2 U. \tag{5}$$

In this equation the matrices H_1 and H_2 relate to the—unknown—system matrices $\{A_1, B_1, \ldots, A_K, B_K\}$ and ditto unknown weights $\{w_{kl}\}$ as

$$\text{vec}(H_1) = W \cdot \text{vec}(A), \tag{6}$$
$$\text{vec}(H_2) = W \cdot \text{vec}(B). \tag{7}$$

The matrices A, B, and W are composed as follows:

$$A = \begin{pmatrix} A_1 \\ \vdots \\ A_K \end{pmatrix}, \quad B = \begin{pmatrix} B_1 \\ \vdots \\ B_K \end{pmatrix}, \quad W = w \otimes I_{N^2} = \begin{pmatrix} w_{1,1} I_{N^2} & \cdots & w_{1,K} I_{N^2} \\ \vdots & \vdots & \vdots \\ w_{M,1} I_{N^2} & \cdots & w_{M,K} I_{N^2} \end{pmatrix}, \tag{8}$$

where $\otimes$ is the Kronecker-product, and I_{N^2} is the $N^2 \times N^2$ identity matrix. Note that Equation (5) is not anymore a linear problem, as the unknown matrices A, B, and W appear in a non-linear way in the equation. This equation is exactly of the type of simple linear networks as in [14]. Therefore, its solution method is fully applicable, so that an efficient and accurate algorithm is available for solving this problem in terms of H_1 and H_2. However, the problem has now shifted to solving two additional non-linear equations:

$$W \diamond A = H_1, \tag{9}$$
$$W \diamond B = H_2. \tag{10}$$

where A, B, and W have to be solved from the known—i.e., computed—matrices H_1 and H_2. The operation $\diamond$ makes the relations in Equations (6) and (7) explicit. This is an underdetermined set of equations that can only be solved by additional information, such as assuming sparsity for A, and a block structure for W, as defined in Equation (4).

3.3 Identification of PWL Models with *Unknown* Switching and *Regular* Sampling from *Poor* Empirical Data

We will now focus on the general case, that the genome wide expressions X, their derivatives $\dot{X}$, and the external inputs U are available as empirical data $\mathcal{D}$. In this case, the objective of system identification is to compute concurrently the system parameters

A, B, and weights *W* (and hence the switching times *S*). Equation (5) provides us with the general state space equations for a PWL system.

In practical experimental conditions, white process and measuring noise adds to the right-hand side. The fit between the empirical data and the system model can be quantified by the weighted difference between observed and expected expression profiles expressed as a linear L_p-criterion:

$$\mathcal{E}_{sys}(A, B, w|\mathcal{D}) = \sum_{k,l} w_{kl} \|A_l x[k] + B_l u[k] - \dot{x}[k]\|_p \tag{11}$$

Here, (A, B) represent the set of system parameters, and $\mathcal{D} \equiv \{X, U, \dot{X}\}$ the observed data, i.e., the measured genome-wide expressions X, their fluxes $\dot{X}$, and the external inputs U. The criterion furthermore involves the relation between the k-th observation and the ℓ-th subsystem Σ_ℓ; namely the *weight* $w_{k\ell}$ and the *distance* $d_{k\ell}$ between observed and the expected value of observation k relative to subsystem model Σ_ℓ.

In order to handle the underdetermined character of the problem, we furthermore employ the sparsity and the hierarchy of the underlying biology. This means that the matrices A_ℓ and B_ℓ are *row-sparse*, but not necessarily collum-sparse, as some genes—called the master-genes or source-genes—control a large part of the entire genome. Under a wide range of conditions, this problem is equal to minimization of the L_1-norm of the rows of A_ℓ and B_ℓ as argued by J. J. Fuchs [7]. This implies a global minimization such as

$$\mathcal{E}_{sparse}(A|\mathcal{D}) = \sum_{\ell} \|\text{vec}(A_\ell^T)\|_1 \equiv \|A\|_1 \tag{12}$$

under the constraints that $\{X, U, \dot{X}\}$ satisfy Equation (5).

The problem of estimating the system parameters can thus formally be defined as the search for the vectors A^*, B^* and w^* that globally minimize $\mathcal{E}$. This can be formulated as a quadratic programming problem, as follows:

QP: given the data $\mathcal{D}$, compute the system matrices A, B and the weight matrix w:

$$(A^*, B^*, w^*) = \arg\min_{(A,B)\in\mathbb{R}^{N(P+N)}, w\in\mathbb{R}^{KM}} \mathcal{E}(A, B, w|\mathcal{D}) \tag{13}$$
   subject to:
$$\mathcal{E}(A, B|\mathcal{D}) = \lambda_1 \mathcal{E}_{sys}(A, B, w|\mathcal{D}) + \lambda_2 \mathcal{E}_{sparse}(A|\mathcal{D}) + \lambda_3 \mathcal{E}_{sparse}(B|\mathcal{D}),$$
$$C_{MK}(w).$$

for selected λ's with: $\lambda_1 + \lambda_2 + \lambda_3 = 1$, and the constraints $C_{MK}(w)$ in Equation (4). This is a regularized (or scalarized) convex quadratic optimization problem that is not well posed because it has a nonsingular Jacobian at the optimum, and becomes ill-conditioned as the iterates approach optimality. Instead of this quadratic programming problem we will therefore study the following two coupled linear programming problems associated to the original QP:

LP1: given the weight matrix $\tilde{w}$ compute the system matrices (A^*, B^*):

$$(A^*, B^*) = \arg\min_{(A,B)\in\mathbb{R}^{N(P+N)}} \mathcal{E}(A, B, \tilde{w}|\mathcal{D}) \tag{14}$$
   subject to:
$$\mathcal{E}(A, B|\mathcal{D}) = \lambda_1 \mathcal{E}_{sys}(A, B, w|\mathcal{D}) + \lambda_2 \mathcal{E}_{sparse}(A|\mathcal{D}) + \lambda_3 \mathcal{E}_{sparse}(B|\mathcal{D})$$

LP2: given system matrices $(\tilde{A}_\ell, \tilde{B}_\ell)$ apply the L_1-norm $\tilde{d}_{kl} = \|\dot{x}[k] - \tilde{A}_\ell x[k] - \tilde{B}_\ell u[k]\|_1$ to compute the weight matrix w^*:

$$w^* = \arg\min_{w \in \mathbb{R}^{KM}} \mathcal{E}_{sys}(\tilde{A}, \tilde{B}, w | \mathcal{D}) = \sum_{k=1}^{M-1} \sum_{l=1}^{K} \tilde{d}_{kl} w_{kl} \tag{15}$$
$$\text{subject to:}$$
$$C_{MK}(w).$$

LP1 is a regularized optimization. J. J. Fuchs [6,7] has described conditions under which the regularization drives the optimization problem towards the global solution. Though these conditions do not strictly apply here, we find that this approach succeeds in numerical simulations. Both LP-problems can be solved efficiently with a partial dual simplex method as in [14], or by using large-scale or interior-points methods. The algorithm to estimate the system parameters $\{A, B\}$ and w consists of iteratively solving the two optimizations LP1 and LP2 subsequently, until the criterion has sufficiently converged. Though the solution of the original quadratic programming problem QP in Equation (13) is also the global solution of the two coupled LP-problems LP1 and LP2, there can also exist local solutions to the couple {LP1,LP2}, unfortunately.

3.4   Construction and Control of the Subsystem Weightmatrix

For small values of the regularization terms in $\mathcal{E}$ in LP1 (Equation (14)), i.e., $\lambda_2, \lambda_3 \ll \lambda_1$, and a simultaneous, extreme under-determined system, i.e., $\#\Sigma_\ell \ll N$, the tandem {LP1,LP2} proposed above, runs into problems. The problem amounts to the degree of freedom that formulation LP1 offers to match empirical data $\mathcal{D}$ with system $\Sigma = (A, B)$ in order to minimize the distance to the model space $d(\mathcal{D}, \Sigma)$. It is well-known that at least $M_{min} \propto log(N)$ measurements are required for a good reconstruction of sparse matrices A and B, see for instance [6,7,24]. Therefore, when $\#\Sigma_\ell \ll M_{min}$, the heavily under-determined system has a high degree of freedom to match the data with the model. This will cause the tandem {LP1,LP2} to halt as the criterion $d(\mathcal{D}, \Sigma) \approx 0$ has been reached.

Avoiding this problem requires (i) the restriction of the maximum number of subsystems to $K < M/log(N)$, and (ii) the careful control of the weight matrix w during the iteration, such that each subsystem Σ_ℓ has at least M_{min} elements, i.e., $\#\Sigma_\ell \geq M_{min}$. For this reason, the following iteration is performed for initializing the weight matrix:

1. Assign the *current measurement* k to 1, and the *current system* ℓ to 1. Initianlize w to the $M \times K$ null matrix: $w = 0$.
2. The first M_{min} measurements are assigned to the current—i.e., first—subsystem: $w(11 = 1, \ldots, w_{M_{min},1} = 1$. Now the current measurement k is set to $M_{min} + 1$.
3. The current measurement, $\xi_k = (x[k], u[k], \dot{x}[k])$, belongs to the current subsystem Σ_ℓ if $d(\xi_k, \Sigma_j)$ is minimized by $_J = \ell$. In that case: (i) it is assigned to the current system by setting $w_{k\ell} = 1$, and (ii) the next measurement is considered, i.e., k is increased, and step 3 is repeated.
4. If another system Σ_j is closer to ξ_k, then this system is assigned to the current system: $\ell = _J$, and measurement k is considered as the first of M_{min} measurements assigned directly to this subsystem, i.e., $w_{k\ell} = 1, \ldots, w_{k+M_{min}-1,\ell} = 1$, k is set to $k + M_{min}$, and step 3 is repeated.

This iteration process is continued as long as there are unassigned measurements. When the final subsystem has less then M_{min} elements, these are discarded. Finally, all measurements will belong to some subsystem, while w obeys all constraints defined in Equation (4). One of the advantages of this *matching* algorithm is that it requires no advance knowledge of the number of subsystems.

3.5 A Tandem for Network Reconstruction Using the Subsystem Weight Matrix

The procedure for constructing and managing the subsystem weight matrix w, defined in Section 3.4, allows for an efficient tandem approach to solving the identification problem.

The non-linear problem $\dot{X} = H_1 X + H_2 U$, defined in Equation (5), can be solved in terms of H_1 and H_2, but not in terms of A, B, and W. It is a bilinear problem in terms of A and B for fixed W, otherwise it is a not well-posed quadratic problem. For these reasons, we again split the problem and follow a tandem approach as discussed in Section 3.2. However, in the present tandem the construction of the subsystem weight matrix w is performed by the matching approach defined above, rather than by the LP2 defined in Equation (15). Both amount to a solution obeying the weight constraints in Equation (4), but the matching algorithm will prevent too underdetermined systems that will prematurely halt the iteration as they generate a fictitious match with the model. The computation of the system matrices (A, B) is again performed by the robust L_1 identification in LP1, with $\lambda_1 = 0$, and $\lambda_2 = \lambda_3$. The tandem is controlled by the distance between the data and the model: $d(\mathcal{D}, \Sigma)) = \mathcal{E}_{sys}(A, B, W|\mathcal{D})$, defined in Equation (11). If this quantity has converged below a pre-specified threshold, the iteration is terminated.

4   Numerical Experiments and Performance of the Approach

The approach described in the previous section resulted in an efficient and fast algorithm that is able to estimate accurately the gene-gene coupling matrix based on several genome-wide measurements, and that is robust towards measurement noise.

All experiments were performed on a PC with an Intel Pentium M processor of 1.73 GHz and 1 GB RAM memory under Windows XP Professional, using Matlab 6.5 Release 13 including the Optimization Toolbox. The latter's routine linprog was used to solve LP problems; its default solution method is a primal-dual interior point method, but an active set method can optionally be used, too. For larger problems, it turned out to be essential for obtaining reasonable computation times that the LP problems were solved by application of the active set method on the dual problem formulation. Therefore, this method was adopted throughout all the experiments.

Since results can depend on the particularities of given data and the original system that generated it, all experiments have been performed on a number of independent runs on randomly selected data and systems. Hence they convey the behavior of our approach "on average". The number of independent runs is 50 for each of the experiments described below.

In line with the definitions above, we use the parameters N, M, K to quantify the size and complexity of the input. In addition, the sparsity of the local interaction matrix A is measured by the number of non-zero entries per row and denoted by k (which should be much smaller than N). To complete the system's data set, some stochastic Gaussian white noise is added to the input data set. It is normally distributed with zero mean and some standard deviation σ that determines the noise level. To quantify the quality of the resulting approximation A_{est} of A^*, a performance measure is introduced: the number of errors N_e.

These errors are generated in the reconstruction by the failure of the algorithm to identify the true non-zero elements of the original sparse matrix A^*. These errors stem from false positives and false negatives in the reconstructed matrix A_{est}. Their numbers are added up to produce the total number of errors N_e.

The success of the algorithm depends on different factors. First, for a certain number of genes, a sufficient number of measurements has to be available. Therefore, the minimal number of measurements required for a certain number of genes, denoted by M_{min}, has been determined. This is the number of measurements so that the total system error, N_e, is acceptably small. Figure 1 represents the values for M_{min} as a function of the number of genes.

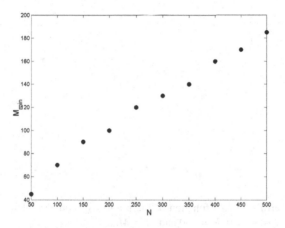

Fig. 1. Minimal number of required measurements M_{min} as a function of the number of genes N

For comparability reasons, the number of genes in all the following experiments has been fixed to $N = 150$. Consequently, the associated minimal number of measurements has been fixed to $M_{min} = 90$ (see Figure 1).

Second, the number of errors N_e depends on the noise level σ. Figure 2 shows how this noise level influences the error rate in our approach. As to be expected, the error increases if the noise level increases, and vice versa.

The numerical experiments consist of the comparison of the reconstructed network with the—known—original network structure, and they clearly reveal the range where the approach is effective.

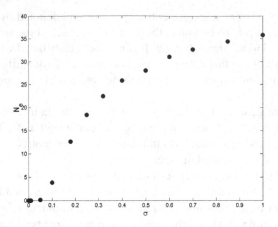

Fig. 2. Number of errors N_e as a function of the noise level σ, with $N = 150$, $M = M_{min}$ and $k = 1$

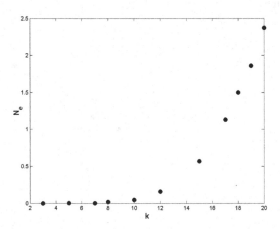

Fig. 3. Number of errors N_e as a function of the number of non-zero elements per row k for a single subsystem ($K = 1$), with $N = 150$ and $M = M_{min}$

A basic assumption in the approach is the sparsity of the underlying gene-gene coupling matrix, represented by the number of non-zero entries per row, k. If k rises above a certain threshold, the performance of the approach is abruptly and severely affected (see Figure 3).

For relatively moderate noise levels and a high degree of sparsity—i.e., a small number k of non-zero elements in the rows of matrix A^*—the approach allows one to reconstruct a sparse matrix with great accuracy from a relative small number of observations $M \ll N$. For example, A^* with rows of 150 components of which all but 3 are equal to zero, can be efficiently reconstructed from just 90 independent measurements (Figure 4).

Figure 4 shows an initial increase, followed by a decrease. Finally, N_e jumps abruptly to zero above a certain threshold value for M. To explain this phenomenon, remember that the number of errors N_e is the sum of the false positives and the false negatives in the gene interaction matrix. The false positives correspond to the non-zero values

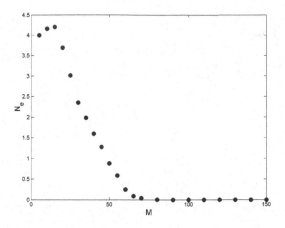

Fig. 4. Number of errors N_e as a function of the number of measurements M, with $N = 150$ and $k = 1$

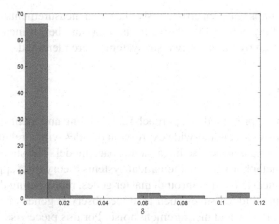

Fig. 5. The distribution of the error measure δ for partioning $M = 200$ measurements into subsystems, with $N = 150$. Two subsystems were identified.

in the matrix A_{est} that should be zero, and vice-versa for the false negatives. Turning back to Figure 4, the initial increase is caused by false positives. Indeed, as long as $M < M'_{min}$, where M'_{min} is the minimal number of required measurements *in the case of a single row*, $k \approx M'_{min}$. As soon as M reaches M'_{min}, the system becomes completely determined, whence k drops to its proper value. Observe that $M'_{min} < M_{min}$ due to the absence of effects related to the composition of rows. Notice that the false negatives decrease monotonously over the entire range of M.

Finally, some experiments concerning multiple subsystems were performed. Figure 5 shows the accuracy of the partioning of the available measurements into different subsystems. The error measure δ shown in Figure 5 is defined as the cumulative distance

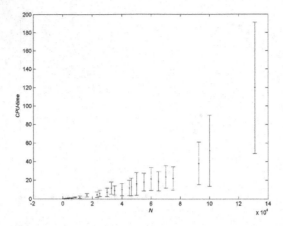

Fig. 6. The CPU time used by the reconstruction algorithm (in seconds) as a function of the number of genes N

in terms of time stamps between erroneously classified measurements and the switching point of the class they belong to, relative to the total number of measurements. In the experiment illustrated by Figure 5, two subsystems were identified.

5 Discussion

In this work, we have presented an approach for modeling and identifying gene regulatory networks from near genome-wide expression profiles with a relative small amount of time instances using a piecewise linear state space model. The state space model is a rich and flexible metaphor from mathematical systems theory that, applied to this case, allows for hierarchical activation through master genes, representing the effects of multiple external inputs, hidden states such as none-observed genes or protein densities, and the effects of process and measurement noise. For this piecewise linear state space modeling, we have presented an identification technique, based on a linear programming problem. This approach resulted in an efficient and fast algorithm that is able to accurately estimate the gene-gene coupling matrix for a large number of genes based on only several hundred genome-wide measurements, and that is robust towards measurement noise. Figure 6 shows the CPU time used by the algorithm as a function of the number of genes N.

In future work, a few difficulties with regard to the system identifiability of this approach, i.e., the potential to reconstruct the interaction network from empirical data, will have to be addressed.

1. Due to the huge costs and efforts involved in the experiments, only a limited number of time points are available in the data. Together with the high dimensionality of the system, this makes the problem severely under-determined.
2. In the time series, many genes exhibit strong correlation in their time-evolution, which is not per se indicative for a strong coupling between these genes, but rather

induced by the over-all dynamics of the ensemble of genes. This can be avoided by persistently exciting inputs.

3. Not all genes are observed in the experiment, and certainly most of the RNAs and proteins are not considered. Therefore, there are many *hidden* states.
4. Effects of stochastic fluctuations on genes with low transcription factors are severe and will obscure their true dependencies.
5. Because the identification techniques work on the rows, the hierarchical principle does not cause a problem, as the gene-gene interaction matrix is highly row-sparse but not column-sparse. In fact, the method utilizes the sparsity of the matrix as an implicit constraint, namely that the value of the components of the matrix should be zero.

With this approach, it is possible to reconstruct the steady states and the associated switching times of a metabolic processes from a set of micro-array experiments. In each steady state the gene-gene interaction matrix defines the network topology. The micro-array technique exhibits a strong increase in efficiency and a simultaneous decrease in associated costs. In the near future, this will enable the registration of large time series of genome wide expression profiles and associated protein densities. The future availability of such data makes the further development of the mathematical modeling and associated identification of dynamic gene expression, such as the approach presented here, an important condition for deducing and understanding the underlying interactions between genes and their environment.

References

1. Bower J.M., Bolouri H.(Editors), Computational Modeling of Genetic and Biochemical Networks, *MIT Press*, 2001.
2. de Jong H., Modeling and Simulation of Genetic RegulatorySystems: A Literature Review, Journal of Computational Biology, 2002, Volume 9, Number 1, pp. 67–103
3. de Jong H., Gouze J.L., Hernandez C., Page M., Sari T., Geiselmann J., Qualitative simulation of genetic regulatory networks usingpiecewise-linear models, Bull Math Biol. 2004 Mar;66(2): pp 301–40.
4. Elowitz M.B., Levine A.J., Siggia E.D., Swain P.S., Stochastic gene expression in a single cell, *Science*, vol.**297**, August 16, 2002, pp.1183–1186.
5. Endy, D, Brent, R. (2001) Modeling Cellular Behavior, *Nature* 2001 Jan 18; 409(6818):391-5.
6. Fuchs J.J. (2003), More on sparse representations in arbitrary bases, in: Proc. 13th IFAC Symp. on System Identification, Sysid 2003, Rotterdam, The Netherlands, August 27-29, 2003, pp. 1357–1362.
7. Fuchs J.J. (2004), On sparse representations in arbitrary redundant bases, IEEE Trans. on IT, June 2004.
8. Glass L., Kauffman S.A. (1973), The Logical Analysis of Continuous Non-linear Biochemical Control Networks, *J.Theor.Biol.*, 1973 Vol. 39(1), pp. 103–129
9. Goldbeter A (2002) Computational approaches to cellular rhythms. Nature 420, 238-45
10. Gonze D, Halloy J, and Goldbeter A (2004) Stochastic models for circadian oscillations : Emergence of a biological rhythm. *Int J Quantum Chem* **98**, pp 228–238.
11. Hasty J., McMillen D., Isaacs F., Collins J. J., (2001), Computational studies of gene regulatory networks: in numero molecular biology,*Nature Reviews Genetics*, vol. 2, no. 4, pp. 268–279, 2001.

12. Novak B, Tyson JJ (1997) Modeling the control of DNA replication in fission yeast, PNAS, USA, Vol. 94, pp. 9147-9152, August 1997.
13. Leiv Øyehaug, Erik Plahte, Stig W. Omholt, Targeted reduction of complex models with time scale hierarchy–a case study, *Mathematical Biosciences*, 185, 123-152, 2003.
14. Peeters R.L.M., Westra R.L., On the identification of sparse gene regulatory networks, *Proc. of the 16th Intern. Symp. on Mathematical Theory of Networks and Systems* (MTNS2004) Leuven, Belgium July 5-9, 2004
15. Plahte E, Mestl T, Omholt SW, A methodological basis for description and analysis of systems with complex switch-like interactions, *Journal of Mathematical Biology*, 36, 321-348, 1998.
16. Rosenfeld N, Young JW, Alon U, Swain PS, Elowitz MB, Gene regulation at the single-cell level, *Science* 307 (2005) pp 1962.
17. Somogyi R., Fuhrman S., Askenazi M., Wuensche A. (1997). The Gene Expression Matrix: Towards the Extraction of Genetic Network Architectures. Nonlinear Analysis, *Proc. of Second World Cong. of Nonlinear Analysis* (WCNA96) 30(3) pp 1815–1824.
18. Swain P.S., Efficient attenuation of stochasticity in gene expression through posttranscriptional control, J Mol Biol 344 (2004) pp 965.
19. Swain P.S., Elowitz MB, Siggia ED, Intrinsic and extrinsic contributions to stochasticity in gene expression, *PNAS* 99 (2002) pp 12795.
20. Steuer R. (2004), Effects of stochasticity in models of the cell cycle:from quantized cycle times to noise-induced oscillations, Journal of Theoretical Biology 228 (2004) 293-301.
21. van Schuppen J.H. (2004), System theory of rational positive systems for cell reaction networks, CWI Report MAS-E0421, December 2004, ISSN 1386-3703
22. Verdult V., Verhaegen M., Subspace Identification of Piecewise Linear Systems, In *Proc. 43rd IEEE Conference on Decision and Control (CDC)*, pp 3838–3843, Atlantis, Paradise Island, Bahamas, December 2004.
23. Westra R.L.,(2005a), Piecewise Linear Dynamic Modeling and Identification of Gene-Protein Interaction Networks, Nisis/JCB Workshop reverse engineering, Jena, June 10, 2005.
24. Yeung M.K.S., Tegnér J., Collins J.J., Reverse engineering gene networks using singular value decomposition and robust regression, *Proc. Nat. Acad. Science*, vol. **99**, no. 9, 2002, pp. 6163–6168.

Sparse Gene Regulatory Network Identification

R.L.M. Peeters and S. Zeemering

Department of Mathematics, Universiteit Maastricht,
P.O. Box 616, 6200 MD Maastricht, The Netherlands
ralf.peeters@math.unimaas.nl, zeemering@math.unimaas.nl

Abstract. In this paper a novel method is presented for the identification of sparse dynamical interaction networks, such as gene regulatory networks. This method uses mixed L_2/L_1 minimization: nonlinear least squares optimization to achieve an optimal fit between the model in state space form and the data, and L_1-minimization of the parameter vector to find the sparsest such model possible. In this approach, in contrast to previous research, the dynamical aspects of the model are taken into account, which gives rise to a nonlinear estimation problem. The set-up allows for the identification of structured or partially sparse models, so that available prior knowledge on interactions can be incorporated. To investigate the potential for applications, the algorithm is tested on artificial gene regulatory networks.

1 Introduction

The number of nonzero parameters in a model determines the *sparsity* of that model, which may be defined as the proportion of zero parameters among the total number of model parameters. The question of how to obtain an accurate sparse model is relevant for many applications. In this paper we investigate an application in *systems biology*: to determine the dominant interactions between a large number of genes. See [12]. It becomes especially relevant to take sparsity into account at an early stage of the system identification procedure in situations where only a limited amount of input-output data is available, possibly of relatively low quality (due to high noise levels, limited opportunity to carry out experiments, high costs involved, etc.).

In this paper an approach to sparse system identification is advocated which employs an L_2-norm to optimize the fit between a model and the data (using a conventional least squares criterion with respect to the vector of prediction errors) and an L_1-norm to minimize the size of the parameter vector to achieve model sparsity. The set of models under consideration is that of state space models in innovations form, using either a full parameterization or a structured parameterization. The sparse system identification procedure allows one to deal with identifiability problems and parameter redundancy, for instance due to a very limited amount of available measurement data. The approach is motivated by applications in the area of reverse engineering of *gene regulatory networks* and inspired by a similar technique described in a static linear setting in the work of [26]. See also [4] and [14] and the

K. Tuyls et al. (Eds.): KDECB 2006, LNBI 4366, pp. 171–182, 2007.
© Springer-Verlag Berlin Heidelberg 2007

references mentioned there. Note that in such applications it typically holds that the number of genes is large (hundreds to thousands), the interaction network is sparse (a gene will be directly influenced by at most 10 to 20 other genes; when it comes to dominant influences this number is even less), the amount of gene expression data obtained from micro-array experiments is small (although the gene expression levels are measured for many genes, the number of time instants involved is usually less than 20), the data are quite noisy (with noise levels up to 30 percent), and the underlying process is in reality nonlinear exhibiting various modes or equilibrium states. However, the situation is currently rapidly improving by the development of new measurement technology, which makes it worthwhile to anticipate such progress and to develop appropriate estimation techniques.

2 Problem Definition

The *model class* considered here is the class of discrete-time linear time-invariant (LTI) *state-space models in innovations form*, described by the equations

$$x[k+1] = Ax[k] + Bu[k] + Ke[k], \tag{1}$$
$$y[k] = Cx[k] + Du[k] + e[k]. \tag{2}$$

Here, at each time instant $k \in \mathbb{Z}$, the n-vector $x[k]$ denotes the state, the m-vector $u[k]$ denotes the exogenous input, the p-vector $e[k]$ denotes the noise input, and the p-vector $y[k]$ denotes the output. It is assumed that a record of input-output observations is available with respect to the exogenous input signal $\{u[k]\}$ and the output signal $\{y[k]\}$. This i/o data record can be used to identify the state-space matrices (A, B, C, D, K). The noise input $\{e[k]\}$ is assumed to constitute a zero mean white noise stationary process with constant covariance $\Sigma > 0$; this is the innovations process from which the model class derives its name. As usual, it is further assumed that: (i) minimality holds; (ii) A is asymptotically stable; (iii) $A - KC$ is asymptotically stable; (iv) for each k the innovation $e[k]$ is independent from the state $x[\ell]$ and the input $u[\ell]$ at time instants $\ell \leq k$, and also from the output $y[\ell]$ at time instants $\ell < k$. For a further background on this model class and its use in system identification, see [16], [11] and [21]; see also [15] and [17].

When modelling gene regulatory networks we shall restrict the model class to the form

$$x[k+1] = Ax[k] + Bu[k], \tag{3}$$
$$y[k] = Cx[k] + e[k] \tag{4}$$

with $K = 0$, $D = 0$, $C = I$ (the identity matrix), and $x[k]$ representing the gene expressions. Then A denotes the gene interaction matrix, $y[k]$ the measured gene expressions and $u[k]$ the input to the microarray experiment. See [26] and [20] for similar model classes. In this model class process noise is disregarded, but measurement noise is taken into account.

Although the focus in this paper is on the identification of sparse gene regulatory networks with the model class (3)-(4), the developed system identification method is applicable to the general class. In the following sections we will therefore describe the method for this general class (1)-(2). We will return to the class of sparse gene regulatory networks in the experiments.

3   Technical Background

To identify the system (A, B, C, D, K) and the noise covariance Σ from an available record of i/o data, we consider the class of *prediction error methods* (PEM). The one-step-ahead predictor associated with an estimate $(\hat{A}, \hat{B}, \hat{C}, \hat{D}, \hat{K})$ of (A, B, C, D, K) is given by the equations

$$\hat{x}[k+1] = (\hat{A} - \hat{K}\hat{C})\hat{x}[k] + (\hat{B} - \hat{K}\hat{D})u[k] + \hat{K}y[k], \tag{5}$$
$$\hat{y}[k] = \hat{C}\hat{x}[k] + \hat{D}u[k]. \tag{6}$$

It gives rise to a *prediction error process* $\{\hat{e}[k]\}$ according to

$$\hat{e}[k] = \hat{y}[k] - y[k] = \hat{C}\hat{x}[k] + \hat{D}u[k] - y[k]. \tag{7}$$

In a practical situation (A, B, C, D, K) is estimated by minimization of the nonlinear least squares criterion of fit

$$V = \frac{1}{M} \sum_{k=1}^{M} \|\hat{e}[k]\|_2^2 \tag{8}$$

over the set of matrices $(\hat{A}, \hat{B}, \hat{C}, \hat{D}, \hat{K})$, where M denotes the size of the data record. It is known that for $M \to \infty$ the unique global minimum value of V occurs at $(\hat{A}, \hat{B}, \hat{C}, \hat{D}, \hat{K}) = (A, B, C, D, K)$ and at systems that cannot be distinguished from (A, B, C, D, K) from an input-output point of view. One important potential drawback of the approach is that the prediction error criterion V is notorious for the many local minima it may possess, especially when noise is present and the size M of the data record is small. When an iterative local search method is employed to minimize V, this makes it necessary either to come up with a good initial estimate (such as may be provided by subspace identification methods; cf. [1], [2], [13], [18], [23]) or to use a sufficiently large number of different initial estimates.

Depending on the chosen parameterization of the model class, there are several factors which contribute to identifiability problems that may arise in the prediction error identification framework described above. One well known possible source of unidentifiability involves state-space basis tranformations, as they do not affect the input-output behaviour of a model. This phenomenon is well understood and can be handled for instance by using canonical forms. In a situation where the number of measurements M is too small, or when the input signals are not sufficiently exciting, there will be subset of dimension > 0 of models that are all consistent with the data, thus causing unidentifiability. With the current state-of-the-art of data acquisition in genetics, such a situation is in fact typical when modelling gene regulatory networks, see e.g., [22], [26], [20]. In this paper we shall deal with this issue by investigating whether it is possible to use additional prior information about the sparsity of a model to select a relevant model from this class.

In a *static* and entirely *linear* estimation setting, a number of results are available in the literature which show that minimization of an L_1-criterion under certain conditions yields a model which is as sparse as possible (i.e., which has as few nonzero parameters as possible). See, e.g., [3], [7], [8] and the references given there. These ideas have first been introduced in the context of *dynamical* models for reverse engineering of gene regulatory networks in [26] and they have been slightly extended in [20]. However, those approaches are restricted to a linear estimation framework by making simplifications which do not take the dynamics fully into account. It is the purpose of the present paper to investigate the merits of an L_1-minimization technique in the context of prediction error identification of state-space models in innovations form, and gene regulatory networks in particular, in situations where unidentifiability caused by data shortage occurs and where the estimation problem is nonlinear.

4   A Prediction Error Method for Sparse System Identification

The basic idea behind the sparse system identification approach investigated in this paper is to first strive for an optimal fit between the data and the model in the conventional nonlinear least squares sense (as achieved by minimization of the criterion V), and then to search within the set of models that minimize V for a sparse one *by minimizing the L_1-norm of the parameter vector*. The parameterizations of (A, B, C, D, K) considered in this approach concern the class of *structured models*, where a certain selection of entries of (A, B, C, D, K) are given fixed prespecified values, while all the remaining entries of (A, B, C, D, K) are collected in a single parameter vector θ. This includes the case of *full parameterizations* (where no entries have a prespecified value), the case of *structured models* as discussed in [5] (where a selection of entries are fixed to zero), and many well-known *canonical forms* (such as modal forms and companion forms, both in the scalar and in the multivariable situation; here some selected entries are fixed to zero while other entries are fixed to 1). It also includes the kind of models proposed for gene regulatory networks in Section 2. Note that the generality and flexibility of this set-up allows one to directly incorporate available prior knowledge on the presence or absence of specific gene interactions in a convenient way.

At a given system (A, B, C, D, K) corresponding to a parameter vector $\theta \in \mathbb{R}^N$, any Gauss-Newton type local search method requires the computation of the corresponding error vector $\hat{e} = (\hat{e}[1]^T, \hat{e}[2]^T, \ldots, \hat{e}[M]^T)^T$ and of its associated $pM \times N$ Jacobian matrix $\hat{J} = \partial \hat{e}/\partial \theta$. The latter can be achieved with the help of the *sensitivity system* which is obtained by partial differentiation of the equations of the prediction error filter (cf., e.g., [10], [19], [9] and [24]).

The kernel of the Jacobian matrix $\hat{J}$ consists of those directions in the parameter space (at the system at hand) for which the entire error vector $\hat{e}$ does not change in first order approximation. (Note: then V does not change in first order approximation either.) The kernel of $\hat{J}$ is a meaningful concept which can be used in the case of both fully parametrized and structured systems and also in the case of 'poor data' (i.e., with few measurements and/or lack of excitation). The orthogonal complement of the kernel of $\hat{J}$ determines the subspace of the parameter space in which the search directions generated by Gauss-Newton type methods are contained, see [25].

The above observations lead to the following two key elements of the sparse system identification method proposed and investigated in this paper:

(1) The orthogonal complement of the kernel of $\hat{J}$ is locally employed as the subspace in which to select a search direction for improving the nonlinear least squares prediction error criterion V (such as, e.g., achieved by Gauss-Newton type optimization methods).

(2) The kernel of $\hat{J}$ is locally employed as the subspace in which to select a search direction for improving the L_1-norm of the parameter vector θ.

There are many ways to build an actual sparse system identification algorithm from these two key elements. Algorithms may differ with respect to the amount of iterations of these two types, and the combinations and order in which they occur. For instance: one may perform steps in the two subspaces $\ker(\hat{J})$ and $\ker(\hat{J})^{\perp}$ simultaneously at each iteration; one may perform steps in the two subspaces alternatingly; one may first perform optimization of V by only taking steps in subspaces of the type $\ker(\hat{J})^{\perp}$ and then minimize the L_1-norm of θ

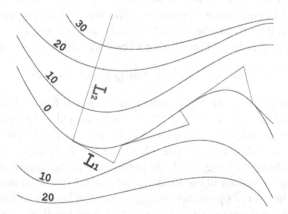

Fig. 1. The iteration scheme: optimization of V in the first step (L_2) and optimization of the L_1-norm of θ in the next (L_1). The contours of the criterion V are shown. It can be seen that a step in the L_1-direction influences the value of V.

afterwards by taking steps in subspaces of the type $\ker(\hat{J})$. In the latter case, the value of V may deteriorate after a number of steps, so that it becomes necessary to incorporate intermediate optimization steps focused on the re-minimization of V, as shown in Figure 1.

Other aspects in which algorithms may differ are: the computation of the search directions in the two subspaces (e.g., depending on the actual Gauss-Newton type method used); the computation of the corresponding step sizes in the computed directions.

5   Mixed L_2/L_1-Minimization

Suppose that an i/o data record is given, which stems from a system with matrices (A, B, C, D, K) and covariance Σ for the innovations process. At a given system $(\hat{A}, \hat{B}, \hat{C}, \hat{D}, \hat{K})$ corresponding to a current parameter vector θ, one may compute the error vector $\hat{e}(\theta)$ by running the associated prediction error filter and the Jacobian matrix $\hat{J}(\theta)$ by running the associated sensitivity system. Then the gradient of the criterion of fit $V = \|\hat{e}\|_2^2/M$ is given up to a scalar factor by $\hat{J}^T\hat{e}$ and the Gauss-Newton (GN) method produces a search direction s given by

$$s = -(\hat{J}^T\hat{J})^{-1}\hat{J}^T\hat{e}. \tag{9}$$

Next, to improve the value of V, the parameter vector θ is modified according to

$$\theta_{\text{new}} = \theta + \alpha s \tag{10}$$

for some step size parameter $\alpha > 0$. In the undamped GN method $\alpha = 1$, but to achieve good convergence behavior it is preferred to determine α by a suitable line minimization procedure (cf. [6]).

Note that $\ker(\hat{J})$ is the space in which the error vector $\hat{e}$ does not change locally around θ in first order approximation. We seek to improve the value of $\|\theta\|_1$ by computing a search direction in $\ker(\hat{J})$, as this will not affect the value of V in first approximation. This leads to the following optimization problem:

$$\text{minimize } \|\theta + s\|_1 \quad \text{subject to: } \hat{J}s = 0 \tag{11}$$

which can be rewritten as an LP problem in standard form. It clearly admits a finite feasible solution and can be solved with standard LP software. If s^* is an optimal solution giving an improved value, then $\|\theta + \beta s^*\|_1$ gives an improved value too for all $0 < \beta < 1$. An actual choice of β should take into account that second and higher order changes in V do not significantly compromise the quality of the fit between the data and the model. It also must be tuned in such a way that convergence of the overall optimization algorithm can be guaranteed to a point for which $\|\theta\|_1$ is minimal among the set of points for which V is (locally) minimal.

One practical heuristic way by which one may attempt to achieve this, is to restrict the maximal relative change in the value of V that is allowed to occur when a value for β is chosen. However, it is not easy to choose an appropriate bound which guarantees monotonic convergence: several experiments have been carried out which exhibit chaotic iteration behavior or cyclic behavior near a local optimum value. For a bound that is not restrictive enough, it has been witnessed that an *increase* of $\|\theta\|_1$ may happen instead of a decrease, resulting as the net effect of an optimization step with respect to $\|\theta\|_1$ followed by an optimization step with respect to V. Also, early in the iteration process the step s^* for $\beta = 1$ may sometimes be large and produce so many zero entries in the matrices (A, B, C, D, K) that non-minimality occurs. This once again makes clear that the choice of β should be treated with care.

6   Experiments and Results

The experiments are focused on identifying artificially generated sparse gene networks, selected from the model class described in Section 2. In this model class only A and B are considered sparse, C is the identity matrix (the output y concerns state observations) and D and K are zero. The original data generating system however, is assumed to be a *continuous-time system*. This means that to perform experiments, the continuous-time system (A_c, B_c, C_c), with A_c and B_c sparse, has to be converted to a discrete-time system (A_d, B_d, C_d), with a certain sampling time T. For small T the discrete-time matrix A_d can be approximated by the following first order approximation:

$$A_d = e^{A_c T} \approx I + A_c T.$$

The exogenous input for the converted system is chosen to be a uniform random signal. The algorithm is tested on situations with Gaussian noise. The initial estimate is either the true system or close to the true system (and in the model class). To find an optimal solution the tested algorithm first uses nonlinear least squares optimization to converge to a (local) minimum of V and then alternates between L_1-optimization of θ (1 step) to improve sparsity and nonlinear least squares optimization of V (10 steps). Computations are performed in MATLAB.

   Previous experiments on small networks with sufficient data have shown that the 2-step optimization algorithm produces appropriate sparse models. However, convergence of the algorithm to a local minimum of $\|\theta\|_1$ on a manifold where V attains a minimum value, appears to be slow. This is not surprising since the proposed minimization technique for $\|\theta\|_1$ employs the tangent space to the manifold and is in essence a gradient technique. The experiments on these small networks have been extended to larger networks with few data, in this particular case to sparse artificial gene networks. The general structure chosen for these gene networks and the structure of the interaction matrix A_c are shown in Figure 2. This structure provides a certain amount of sparsity that increases with the number of genes (Table 1).

   The models included in the experiments are networks with $n = 10$ or $n = 20$ genes, 1 input, 10 or 20 outputs and Gaussian measurement noise. The initial estimate is chosen to correspond to the underlying data generating system.

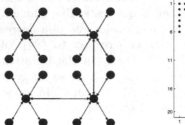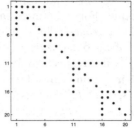

Fig. 2. The structure of a network of 20 genes (left) and the corresponding structure of the interaction matrix A_c (right). The structure consists of groups of 5 genes. Each group has a central gene that is connected to the 4 other genes and to the central gene of the neighboring group. Each gene is connected to itself. All connections are made in both directions.

Table 1. The number of genes and the sparsity of the matrix A_c using the structure described in Figure 2

Number of genes	Total entries in A_c	Non-zero entries in A_c	Sparsity of A_c
5	25	13	48%
10	100	28	72%
20	400	58	85.5%
40	1600	118	92.63%
100	10000	298	97.02%
⋮	⋮	⋮	⋮

The number of artificially generated gene expressions measurements per gene is $M = 5$ in the case of 10 genes and $M = 10$ in the case of 20 genes. Two important insights emerged from these experiments:

1. In the initial phase of the estimation procedure when V is first minimized, the initial value of $\|\theta\|_1$ may change dramatically.
2. The generated gene expression trajectories are often highly correlated, which makes it difficult to determine the interactions of certain genes.

The first aspect is illustrated in Figure 3. The startup procedure minimizes V without taking the value of $\|\theta\|_1$ into account. Note that the initial value which corresponds to the underlying data generating system is likely not to yield a minimum of V, due to lack of measurements and the presence of noise. Instead, the parameters in matrix A and B are adjusted to account for these deviations. Putting emphasis on the minimization of V apparently leads to a (significantly) higher value of $\|\theta\|_1$. Despite a sparse initial guess, the number of iterations required to subsequently minimize the value of $\|\theta\|_1$ is large.

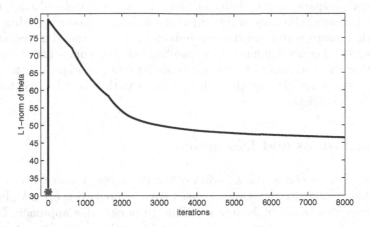

Fig. 3. The startup procedure leads to an increase in the value of $\|\theta\|_1$. (The marker * denotes the the L_1-norm of the original solution.) The number of genes in this example is 50.

The second problem is of a more fundamental nature. The small number of measurements for each gene can lead to highly correlated gene expression profiles, as illustrated in Figure 4. The cause of this problem lies of course in the limited number of measurements, but also in the structure of the model and the characteristics of the input signal. The result is that a certain gene trajectory (the series of measured gene expressions) is often linearly proportional to an other gene trajectory. The fact that the algorithm favours sparse models (and

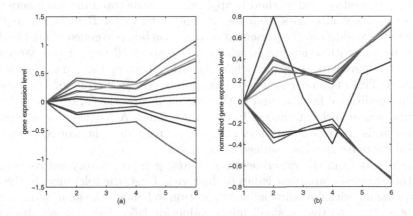

Fig. 4. (a): Gene expression trajectories of length 5 (when excluding the zero inital state) generated by a 10 gene network structured as in Figure 2. (b): Normalized gene expression trajectories (all trajectories are scaled to length 1). It can be seen that there are two distinct groups of highly correlated trajectories.

consequently a sparse matrix A) means that among the correlated genes the gene with the largest excitations in its trajectory will have a corresponding column in A with nonzero entries on the row indices of the correlated genes, while the other correlated genes will have corresponding columns consisting of zeros. This implies that the interactions that these genes have in the original system cannot be properly distinguished as they are lumped together in a limited number of interaction coefficients.

7   Conclusions and Discussion

The strategy to minimize the L_1-norm of the parameter vector θ over the set of models which minimize the nonlinear least squares prediction error V, has interesting properties which make it worthwhile to pursue this approach. For large sparse models with poor data, previous experiments in the literature have shown in a linear estimation setting that sparse models can be quite well retrieved. In the current approach, the dynamical aspects are properly taken into account, in contrast to previous work where simplifications are made to arrive at a linear estimation problem.

Computation of search directions for L_1-norm optimization can be performed effectively and fast, as this involves the solution of an LP problem. The same holds for numerical optimization of the nonlinear least squares criterion V with a Gauss-Newton type iterative local search method. There are however some convergence issues to be dealt with when the two optimization problems are combined. Nevertheless, the method in the form presented here can already be used for models consisting of several hundreds of genes, producing results in reasonable time without having to resort to special hardware and software.

The sparse estimation method is applicable to structured models where only a selected subset of entries from the system matrices (A, B, C, D, K) requires estimation. In addition, the presented method has been extended along the lines of [20] to deal with situations in which only a specified part of the parameter vector is required to be sparse while the other part of the parameter is not penalized. This makes it possible to incorporate certain interactions in a model in an unconditional fashion, and in this way to incorporate prior knowledge on the existence of such interaction from other studies. A particularly interesting application is the identification of sparse gene regulatory networks, where the gene interaction matrix is assumed to be sparse.

The results from the experiments on sparse gene regulatory networks show that there are two important issues to deal with. First, the tolerances in the two steps of the algorithm require online adaptation to balance the joint optimization of V and $\|\theta\|_1$, so that a small initial value for $\|\theta\|_1$ does not get discarded completely when minimizing V in the startup phase of the algorithm. Second, and more important, the trajectories of the artificially generated gene expressions for each gene have to be more or less uncorrelated to ensure that the interactions of each gene can be identified. The solution for this second problem could be found by designing more appropriate sparse gene regulatory networks and input signals. Another possibility is that this is a *fundamental problem* that cannot easily be solved. It may be possible to create artificial circumstances in which the (short) gene expression trajectories are sufficiently uncorrelated to identify their individual interactions, but in real life experiments there is no way to enforce this correlation condition. A useful approach to this problem can be to determine which gene expression trajectories are highly correlated and to treat those genes as one single gene (that is to *cluster* those genes) before identifying the network.

References

1. D. BAUER, Subspace algorithms, *Proceedings of the 13th IFAC Symposium on System Identification*, Rotterdam, The Netherlands, pp. 1030–1041, 2003.
2. D. BAUER, Asymptotic Properties of Subspace Estimators, *Automatica* **41***(3)*, Special Issue on Data-Based Modeling and System Identification, pp. 359–376, 2005.
3. P. BLOOMFIELD AND W.L. STEIGER, *Least Absolute Deviations: Theory, Applications, and Algorithms*, Birkhäuser, Boston, 1983.
4. P. D'HAESELEER, S. LIANG AND R. SOMOGYI, Genetic Network Inference: From Co-Expression Clustering to Reverse Engineering, *Bioinformatics* **16***(8)*, pp. 707–726, 2000.
5. J.-M. DION, C. COMMAULT AND J. VAN DER WOUDE, Generic properties and control of linear structured systems: a survey, *Automatica* **39**, pp. 1125–1144, 2003.
6. R. FLETCHER, *Practical Methods of Optimization*, John Wiley and Sons Ltd., Chichester, 1987.
7. J.-J. FUCHS, More on sparse representations in arbitrary bases, *Proceedings of the 13th IFAC Symposium on System Identification*, Rotterdam, The Netherlands, pp. 1357–1362, 2003.
8. J.-J. FUCHS, On sparse representations in arbitrary redundant bases, *IEEE Transactions on Information Theory* **IT-50***(6)*, pp. 1341–1344, 2004.
9. W.S. GRAY AND E.I. VERRIEST, Optimality properties of balanced realizations: Minimum sensitivity, *Proceedings of the 26th IEEE Conference on Decision and Control*, Los Angeles, CA, USA, pp. 124–128, 1987.
10. N.K. GUPTA AND R.K. MEHRA, Computational aspects of maximum likelihood estimation and reduction is sensitivity function calculations, *IEEE Transactions on Automatic Control* **AC-19**, pp. 774–783, 1974.
11. E.J. HANNAN AND M. DEISTLER, *The Statistical Theory of Linear Systems*, John Wiley and Sons, New York, 1988.
12. H. KITANO, Systems Biology: a brief overview, *Science*, **295**, pp. 1662–1664, 2002.
13. W.E. LARIMORE, System identification, reduced order filters and modeling via canonical variate analysis, in: H.S. Rao and P. Dorato (eds.), *Proceedings of the 1983 American Control Conference 2*, Piscataway, NJ, pp. 445–451, 1983.
14. R. LAUBENBACHER, B. STIGLER, A computational algebra approach to the reverse-engineering of gene regulatory networks, *Journal of Theoretical Biology* **229**, pp. 523–537, 2004.
15. L. LJUNG, *MATLAB System Identification Toolbox Users Guide*, Version 6, The Mathworks, 2004.
16. L. LJUNG, *System Identification: Theory for the User* (2nd ed.), Prentice-Hall Inc., Englewood Cliffs, NJ, 1999.

17. B. NINNESS, A. WILLS AND S. GIBSON, The University of Newcastle Identification Toolbox (UNIT), *Proceedings of the 16th IFAC World Congress*, Prague, 2005.
18. P. VAN OVERSCHEE AND B. DE MOOR, *Subspace Identification for Linear Systems*, Kluwer Academic Publishers, 1996.
19. R.L.M. PEETERS, System identification based on Riemannian geometry: theory and algorithms. *Tinbergen Institute Research Series* 64, Thesis Publishers, Amsterdam, 1994.
20. R.L.M. PEETERS AND R.L. WESTRA, On the identification of sparse gene regulatory networks, *Proceedings of the 16th International Symposium on the Mathematical Theory of Networks and Systems*, Leuven, Belgium, 2004.
21. T.S. SÖDERSTRÖM AND P. STOICA, *System Identification*, Prentice-Hall, New York, 1989.
22. J. TEGNÉR, M.K.S. YEUNG, J. HASTY AND J.J. COLLINS, Reverse engineering gene networks: Integrating genetic perturbations with dynamical modeling, *Proceedings of the National Academy of Science* **100***(10)*, pp. 5944–5949, 2003.
23. M. VERHAEGEN, Identification of the deterministic part of MIMO state space models given in innovations form from input-output data, *Automatica* **30**, pp. 61–74, 1994.
24. E.I. VERRIEST AND W.S. GRAY, A geometric approach to the minimum sensitivity design problem, *SIAM Journal on Control and Optimization* **33***(3)*, pp. 863–881, 1995.
25. A. WILLS, B. NINNESS AND S. GIBSON, On Gradient-Based Search for Multivariable System Estimates, *Proceedings of the 16th IFAC World Congress*, Prague, 2005.
26. M.K.S. YEUNG, J. TEGNÉR AND J.J. COLLINS, Reverse engineering gene networks using singular value decomposition and robust regression, *Proceedings of the National Academy of Science* **99***(9)*, pp. 6163–6168, 2002.

Author Index

Lecture Notes in Bioinformatics